"Any Time Is Trinidad Time"

Kevin K. Birth

"Any Time

University Press of Florida

# Is Trinidad Time"

## Social Meanings and Temporal Consciousness

Gainesville · Tallahassee · Tampa · Boca Raton · Pensacola · Orlando · Miami · Jacksonville

04  03  02  01  00  99  6  5  4  3  2  1

LIBRARY OF CONGRESS CATALOGING-IN-PUBLICATION DATA
Birth, Kevin K., 1963–
Any time is Trinidad time: social meanings and temporal consciousness / Kevin K. Birth.
p. cm.
Includes bibliographical references and index.
ISBN 0-8130-1713-0 (cloth: alk. paper)
1. Ethnology—Trinidad. 2. Time—Social aspects—Trinidad.
3. Time perception—Trinidad. 4. Cacao farmers—Trinidad.
5. Trinidad—Social life and customs. I. Title.
GN564.T7B57  1999
304.2'3—DC21  99-20574

The University Press of Florida is the scholarly publishing
agency for the State University System of Florida, comprising
Florida A&M University, Florida Atlantic University, Florida International University, Florida State University, University of Central Florida, University of Florida, University of North Florida,
University of South Florida, and University of West Florida.

University Press of Florida
15 Northwest 15th Street
Gainesville, FL 32611
http://www.upf.com

To my parents

# Contents

# Preface

Trinidadians attribute the phrase "Any time is Trinidad time" to a calypso sung by Lord Kitchener. The phrase has since entered the everyday language. This project explores time as a cultural construction. From this perspective, this Trinidadian idiom implies that any cultural construction of time is a Trinidadian time. Considering the great ethnic and cultural diversity of the island, there is probably a great deal of truth to this implication. The task that I have set for myself is to explore the social consequences of multiple, coexistent cultural constructions of time. Through working out these consequences, I have come to realize that the use of cultural constructions of time is a major means of defining social differences—in particular ethnic, gender, generational, and class differences—in daily life. The manner in which abstract categories, such as class, become enacted in daily life is a long-standing problem in social science. I suggest that I have found one means, albeit not the only one, of dealing with this issue.

Several excellent reviews of the anthropological literature on time have appeared in this decade (Gell 1992; Greenhouse 1996; Munn 1992; Östör 1993), so I do not feel another review is needed. In addition, there has been a change in the anthropological study of time. Formerly, following the work of Durkheim (1965), Hubert (1909), and Mauss (1979), the majority of anthropological discussions of time emphasized superorganic models of culture and debates concerning the cultural relativity of conceptions of time (Greenhouse 1996:2–3). Recent work, exemplified by Greenhouse (1996) and the collection of essays *The Politics of Time* (Rutz 1992), emphasizes how culturally constructed ideas of time are used in social action and the exercise of power.

Studying time is unlike studying many other topics. One cannot simply walk up to someone and say, "Tell me about your conceptions of time." For this reason, I shall outline the methods that I used. When I first arrived in Anamat, the village I intended to study, I performed three tasks: I conducted a village-wide census; I made a map of the village; and I spent a great deal of time trying to discover Trinidadian metaphors and idioms concerning time. My next step was to investigate the ways in which the village had changed since Dr. Morris Freilich's research in the late 1950s (see Freilich 1960, 1961, 1968, 1980; Freilich and Coser 1972). To do this I talked with groups of people and corroborated their reports with published works on Trinidad. At the same time, I collected information on kinship. I also conducted case studies of important events in the eyes of the villagers, gathering information by selecting particular nodes in the village's informational network—places such as rum shops, taxi stands, the recreation ground during cricket games, and places in the nearest market town where people from the village would loiter. With this information, I gained a general understanding of the social organization of the village.

After obtaining this general understanding, I began interviews on time. I selected informants based on ethnicity, occupational history, age, and gender.

The interview sequence covered the following topics, in this order: gathering personal histories, particularly with regard to occupation; discussing Trinidadian idiomatic expressions concerning time to discover when, where, with whom, and how these expressions are appropriately used; exploring case studies based on events I had witnessed where these idiomatic expressions were used; eliciting reactions to and interpretations of American metaphors of time, particularly metaphors of time as a commodity; and discussing the relationship among past, present, and future in terms of personal history and future expectations, village history, national history, and international history.

A key component throughout fieldwork is observation. Differences in times are not always apparent until there is a conflict between individuals.

Initially, I was embroiled in many such conflicts because of my ideas of time conditioned by growing up in suburbs in the northeastern United States. I appreciate the patience and politeness of my Trinidadian friends and their use of strategies to obscure and gloss over our differences rather than any to engender greater conflict. Using my blunders and the encounters I witnessed between Trinidadians, I developed a set of issues and questions to explore with my informants during the interviews.

Relying on observation has implications for ethnographic representation. In this era of concerns about capturing the "voice" of the "Other," it must be remembered that observed events do not have voices, but observers do. Any voice, whether that of informant or ethnographer, recasts physical, visual, and nonlinguistic aural experience into words. Life is not a text, although it can be represented as one. Consequently, many descriptions are my voice, whereas many interpretations and analyses of events arise out of an interplay of my voice and the voices of my informants. As I have written elsewhere (Birth 1990), I view ethnography as essentially referential so that it will be criticized and challenged. While this might lead to "suppressing voices," I hope that voices will emerge to discuss and criticize my work, because I am simply one voice among many contributing to discussions about Trinidad, not the voice to represent Trinidad.

Informants' names and place-names, including Anamat, are all pseudonyms because of a request from Morris Freilich, as well as the attempt to keep the identity of my informants confidential.

What follows is largely the product of chance, good advice, and an attempt to take advantage of the circumstances in which I found myself. While my choice of Trinidad as a research site was deliberate, my choice of a locale in Trinidad resulted from a telephone call to Morris Freilich who, without knowing anything about me except that I was a graduate student at the University of California at San Diego, generously suggested that I study the village in which he conducted his own doctoral research in 1957–58. By chance, while I was reading his dissertation, I was also wading through Kant's *Critique of Pure Reason* (1929) and rereading Taussig's *The Devil and Commodity Fetishism in South America* (1980). Freilich discussed the temporal orientations of the people of the village of Anamat. Kant discussed the a priori temporal structures of thought. Taussig briefly refers to the commodification of time in capitalist systems. Until this point, I had been intending to study religion in Trinidad, but this coincidental convergence of ideas concerning time piqued my interest and made me rethink my entire project.

My research project began as an attempt to test hypotheses derived from the historical literature on the Industrial Revolution and the psychological literature on achievement and future-orientation. Those two literatures complemented one another, leading me to associate the ability to adapt to the penetration of industrial capitalism with having a well-developed future-orientation. The form has long been associated with the Protestant work ethic that Weber (1958) described as the catalyst for European capitalism taking the form it did.

The complexity of the research setting led me to discard not only the hypotheses I set out with but many of the assumptions on which they were based. I discarded the association of a single conception of time with industrial capitalism or with a peasant economic formation. I discovered that people utilize many different conceptions of time and that these are distributed in complex ways over social contexts and statuses. Finally, I became concerned not only with the times that individuals possessed but the fashion in which they deployed these times.

After arriving in Trinidad, my initial dilemma was to find a way to get to Anamat, then find a place to live there. Again, by chance, the chairman of the Department of Government at the University of the West Indies at that time, John G. La Guerre, was from Anamat. Indeed, he expressed gratitude toward Freilich for motivating him to pursue an academic career. Dr. La Guerre arranged accommodations for me in Anamat with his mother, and that is where my wife and I stayed for the first five months of my fieldwork.

My entry into Anamat also benefited from several coincidences. First, I arrived shortly before the 1989 soccer match between Trinidad and Tobago and the United States that determined who would play in the 1990 World Cup tournament. Many of those in Anamat are soccer fans, and I enjoyed many evenings being teased about the slowness of the American team and getting an education in soccer, both on and off the field. Ironically, after my Anamatian friends worked so hard to convince me that Trinidad had the better team (and I think they did), the United States won.

Soon after the soccer match, Christmas season began. For the first time in several years, the local Catholic church wanted its youth group to go Christmas caroling. They had heard that I played guitar and asked me to accompany them. Soon afterward, the Missionary Baptist Church also asked me to play guitar for their Christmas caroling. As a result, while I was conducting a house-to-house census during the day, I was often visiting the same homes in the evening under different, more informal, and more comfortable circumstances. The leadership of both churches opened many doors for me, for which I am grateful.

My guitar playing attracted the attention of several local *parang* bands, which then invited me to join them. I soon discovered that special and strong ties developed between musicians in *parang* bands. My fellow *paranderos* were great musical companions and important sources of information. I thank them for their patience reflected in their tolerance of my persistent questions and of the slowness with which I picked up *parang's* rhythms.

Consequently, over the first several months of my fieldwork, my identity in the community evolved from simply the American student replicating what Morris Freilich had done in the late 1950s to an identity including soccer fan, guitarist, and *parandero*. The people, coincidences, and events that led to the last three roles did far more to encourage rapport than anything I could have done by design or careful planning.

I attribute my ability to take advantage of chance to the advice and training I have received. I have benefited from excellent teachers, both at the University of Rochester and at the University of California at San Diego. Among my teachers, I must include the Trinidadians with whom I worked. This project never would have been possible without the people of Anamat. In large part, this book is a cooperative effort with them. I cannot imagine a better group of people with whom an anthropologist could work. They provided mountains of information, challenged my conclusions, and had high intellectual standards. I hope that I have been a satisfactory student of theirs.

Many individuals worthy of special mention have made contributions to this project.

In developing the research methods with which I studied time, Roy D'Andrade played an important role of a supportive skeptic and critic. Morris Freilich has provided me with important background on Anamat and methodological advice, and he has served as a never-ending source of encouragement. In preparing my initial project, Stephen Glazier and Daniel Crowley also gave me valuable advice, and the reading list on Trinidadian history that Bridget Brereton sent me has been a valuable guide to the historical literature.

Once I arrived in the field John La Guerre served as an important early guide to the conceptual hazards of studying Trinidad, in particular challenging my implicit, distinctly American conceptions of race and ethnicity.

I have received tremendous amounts of constructive criticism and support from David Jordan, Marc Swartz, Fitz John Porter Poole, and Donald Tuzin. Donald Tuzin, in particular, has gone above and beyond the call of duty in reading my many drafts and providing insightful, incisive comments. I also owe a great deal to the nonanthropologists who have introduced me to scholarly work outside the secure boundaries of my discipline. These individuals include Steven Hahn, Robert Ritchie, and George Lipsitz.

Over coffee, Alex Bolyanatz provided many hours of discussions concerning all facets of the book. Susan Love Brown and Mark Cravalho asked important questions in the early stages of writing after I returned from the field. My colleagues at Queens College of the City University of New York have also contributed to this effort. Warren DeBoer and Sara Stinson both read portions of the book and gave insightful criticisms. I would also like to thank Kevin Yelvington, Mark M. Smith, Henry Rutz, and the anonymous reviewers for their comments. Last, Meredith Morris-Babb and Judy Goffman of the University Press of Florida have played crucial roles in helping me through the publishing process.

Of course, what appears here is entirely my responsibility.

This research has been supported by a USIA Fulbright and by two PSC–CUNY grants (#666436 and #666476). Portions of this work appeared as an article in *Anthropological Quarterly* 69(2) (April 1996):79–89 as "Trinidadian Times: Temporal Dependency and Temporal Flexibility on the Margins of Industrial Capitalism."

Finally, a very special thanks is reserved for my wife, Margaret, whose patience, encouragement, conversations, questions, critical eye, and knowledge of Trinidad have played significant roles in the completion of this project.

# Introduction

People use time to measure, to organize, and to manage experience. Anthropological studies show that concepts of time vary from society to society, and within societies (Gell 1992; Greenhouse 1996; Hall 1983; Munn 1992; Östör 1993; Rutz 1992). Historical studies demonstrate that while specialized attempts to measure time are old, the widespread use of clocks to measure time is recent (Borst 1993; Dohrn-van Rossum 1996; Landes 1983). Indeed, clocks are one tool among many used for the temporal organization of experience; other tools include calendars, ideas of seasonal changes, and knowledge of diurnal cycles of plants and animals. The clock has been closely linked to industrial production (Mumford 1963). Other forms of production, such as pastoralism or peasant farming, involve other ways of thinking about time (Bourdieu 1963; Bourdillon 1978; Evans-Pritchard 1940; Glassie 1982; Hallowell 1955, Le Goff 1980; Michael Smith 1982).

The contrast between agrarian rhythms and industrial times has become a common element of studies of time and has led to many important insights about political, economic, and social changes in Europe and North America (Mumford 1963; Thompson 1967). Yet, this contrast also obscures a fundamental similarity in all conceptions of time: the use of concepts of time to think about the cycles and changes in one's surroundings and to organize activity and social relationships. Whether one is talking about the Native American Saulteaux reckoning the ending of winter by the migration of birds (Hallowell 1955:228) or a hospital staff timing rounds and doses of drugs (Zerubavel 1979:22–36), there remains a similarity in how these different groups of people use models of time—both use temporal models to organize experience. In places such as rural Trinidad, where the influences of industrialism and agrarianism are both felt, the use of multiple and varied temporal ideas to organize experience and activity leads to tension and conflict, as well as to strategic choices in how ideas of time are applied. As a result, a rigid contrast between industrial and agrarian times is untenable (Adam 1995; O'Malley 1992), as is a related contrast between Western and non-Western times (Gupta 1994).

Frequently, such contrasts are linked with ideological claims of moral or economic superiority. For instance, the attribution of temporal attitudes forms an important component of stereotypes of laziness, thrift, or discipline. Such ideological claims have probably been examined most closely with regard to the interaction of class relations and racial categories in the United States, beginning with the slave system of the American South (Genovese 1976:285–309; Mark Smith 1997). In effect, where many different cultural traditions and many different forms of economic activity meet, the diversity of times is great, and so are the potential conflicts. Where there is this potential for conflict, ideas of time become contested and employed in ideological struggles about cooperation, hierarchy, group definitions, and the wielding of power. This book examines these issues in rural Trinidad, the southernmost island of the West Indies.

## "Any Day and Any Time"

When I first arrived in Trinidad, I noticed that many rum shops displayed a placard that read "Opening and closing hours: any day and any time." Venturing into one shop, I asked about the meaning and purpose of the sign. The person behind the counter described Trinidad's complicated liquor licensing laws. These laws determine different hours for different categories of liquor licenses. I then asked when, exactly, the shop opened and closed. This brought a smile to the face of the shopkeeper, who said, "Like the sign says, any day and any time. You from America, eh? Somet'ing you mus' learn about we Trinis: We say, 'Any time is Trinidad time.' Well, that is truth. Sometime I open early. Sometime I don' open at all. I open when there are customers. When there are none, I close the shop and take a little five [a little rest]. When they call, I open again."

The sign signified that the shopkeeper opened or closed the shop at any time, based on her activities and schedule. The sign also indicated that customers might show up at any time based on their activities and needs. Indeed, shopkeepers (and their neighbors) complain that in the middle of the night, customers come to the shop bellowing for it to open. The desire for cigarettes usually prompts such nocturnal demands, but a shopkeeper can never be sure what the request will be for: sometimes customers ask for medicine for a sick family member. People with many different daily routines and rhythms use the shop, and the "any time" on the placard reflects this awareness of multiple routines: those of taxi drivers, factory workers, agricultural laborers, forestry workers, schoolchildren, home-makers, and underemployed young adults. Shopkeepers cater to multiple times because their clientele follows different times.

Anthropology describes different means of temporal reckoning found throughout the world (see Munn 1992). This has led to debate over whether this variation reflects culturally variable consciousness of time or whether it reflects ideological uses of temporal ideas separate from daily experiences of time (Bloch 1977, 1979; Bourdillon 1978; Howe 1981). According to Maurice Bloch, ideological uses of time vary cross-culturally. He suggested that the greater the institutionalized hierarchy, the more important ritual time is in justifying the hierarchy ideologically (1977:289). In effect, rather than each society having one, single time, Bloch suggested that societies with institutionalized hierarchies possess ideas of time ideologically reinforcing hierarchy and other ideas of time that are related to everyday affairs. He accused anthropology of focusing on the former ideas and constructing characterizations of societies from them.

In fact, in attempts to provide holistic, consistent portrayals of other societies, anthropology has tended to portray homogeneity rather than heterogeneity (Clifford 1986; Marcus and Fischer 1986). It has also con-trasted "primitive" societies with "civilized" societies. While this contrast has now fallen into disrepute, it was an important component of early examinations of time. These comparisons did raise some interesting points, such as A. Irving Hallowell's suggestion that, among the Saulteaux, "temporal orientation depends upon the recurrence and succession of concrete events in their qualitative aspects," resulting in temporal orientations that are "local" (1955:233, 234). These events include lunar phases and seasonal changes. Hallowell contrasted such local time with clock-determined temporal orientations of "Western civilization." As he pointed out, such orientations are global—all hours are treated as the same, whether measured at the North Pole which has no daylight during the winter, or at the equator where the seasonal changes in the length of the day are nonexistent, or, by implication, in outer space in a spacecraft moving in its orbit at a different pace than the rotation of the earth.

This contrast of global and local remains relevant, as increasing concern with globalization in the social sciences indicates (Appadurai 1996; Hannerz 1992; Miller 1995, 1997; Olwig 1993; Wolf 1982). One of the consequences of globalization is the coexistence and often the conflict between cultural ideas that are emergent locally and those resulting from global ties. Such is true of cultural ideas of time. Just as culture theory has shifted from describing cultures as homogenous to understanding the distribution and interaction of shared ideas (Goodenough 1981; Hannerz 1992; T. Schwartz 1978; Shore 1996; Swartz 1990), the study of cultural ideas of time has shifted from characterizations of a society's time to discussions of multiple times and their relationships (Adam 1995; Greenhouse 1996; Rutz 1992). In a place such as Trinidad, with a long history of diverse populations, global ties, and local innovations, there are multiple activities with multiple times and myriad ideological uses for ideas of time. To understand the social significance of ideas of time in Trinidad requires exploring the dynamic relationship between externally imposed and locally emergent models of time.

### Anamat

The location of this study is Anamat, a community in rural Trinidad. To someone who is not familiar with Anamat, it appears to be a small, sleepy village with narrow, winding roads, along which stand occasional clusters of homes. At the major junction in the village there is a small shop. There are about five such shops in the community. In 1989, the one at the junction was one of several that sold alcoholic beverages—mostly rum and beer—although in 1998 it was the only such shop. Around these small stores, men chat and periodically buy cigarettes and drinks. Across from the shop at the junction is a "trace," or a lane, which leads to the village's recreation ground. Traces vary in size from narrow, muddy tracks to roads paved with asphalt and wide enough for trucks to pass. The recreation ground, when the abundant grass is kept under control, is the venue for Anamat's cricket and soccer teams. It also serves as a place to hold "sports," which are fund-raising occasions consisting of several athletic competitions with small prizes. The trace then continues into the cocoa walks (the groves of cocoa trees). At first glance, Anamat seems to be an isolated, agrarian settlement.

One of the main activities that occurs at and around the shop is liming. Liming is the art of doing nothing with one's peers, until there is something to do; talking about nothing in particular, unless there is something important to discuss; and sharing important information about jobs, members of the opposite sex, fetes, wakes, neighbors, politics, the police, gambling, and drugs—unless it is prudent to keep one's mouth closed (see Brana-Shute 1976, 1979; Eriksen 1990; Lieber 1976; Wilson 1971).

There are several places along the road where men lime. Accessibility to a shop, a place to sit in the shade, and opportunities for quick escape if the police come while those liming engage in petty gambling determine these locations. Liming can begin early in the morning, with the gathering of men awaiting taxis to take them to work included in the lime. On many days, liming continues through the day. Soon after dark, the noise of the men talking along the road begins to quiet, except for the young men (the "boys on the block" as they call themselves) who sit on a particular set of culverts known as "the block." They talk until they decide to walk home. While the men seem to congregate near the shops and along the road, women gather in the front "galleries" (porches) of homes or underneath "upstairs" houses (houses built on platforms).

Anamat rarely catches the attention of the rest of Trinidad. Despite Trinidad's small size, many Trinidadians have no knowledge of Anamat: when I tell them I did research in Anamat, they respond, "Where's that?" Nestled deep within the central mountain range of the island, it is a village in which roads end, rather than crossing on their way elsewhere. The two roads leading into the village, one from the north, and one from the west, wind through the rolling hills of the Central Range. Due to the poor road foundations and heavy rains, in many places the road surface has shifted, or even disappeared, down the sides of hills. In several spots, the roads have been shifting for as long as people can remember. The Ministry of Works and Infrastructure deals with the problem by filling the sunken portions of road with gravel or dirt several times a year. For many years, year-in and year-out, workers have deposited several truckloads of dirt and gravel on these "land slips," yet every year the land slips further, requiring more truckloads of dirt and gravel in order to keep Anamat linked to the outside world. Recently, the government has tried to reinforce the road and patch all of the land slips with asphalt, but Anamat is still marked by a land slip so severe that engineers shrug their shoulders in helplessness and comment that the road never should have been built there to begin with. Anamatians use this land slip to mark one of the boundaries of the village.

The poor quality of the roads discourages outsiders from coming to Anamat, but the roads do not prevent Anamatians from leaving. Many commute to jobs outside the village on a daily basis, and almost everyone travels to the nearest market town, on occasion, to conduct business. Thus, while Anamat is a place where roads end, its residents are closely tied economically and socially to the world outside the village.

Contemporary Anamat contains a tantalizing diversity of groups. With regard to ethnicity, Anamat's history reveals how many distinct groups—French, Venezuelan, Amerindian, Indian, and African Creole—have been molded into only two ethnic groups: the Creoles and the Indians. Labor relations and patronage have been instrumental in the evolution of these two categories. The cultural construction of the categories of Creole and Indian is an ongoing process. As will be shown in subsequent chapters, the coordination of social activities according to different models of time, or the lack of coordination, becomes a conceptual resource in the process of maintaining or modifying these ethnic categories.

Examining cultural ideas of time in a place that is neither purely agrarian nor purely industrial poses challenges to the conceptual framework that has dominated social scientific and historical approaches to time. The contrast between industry and agriculture underlies this perspective. Industrial time is linear, divisible, measured by clocks, and easily turned into a commodity when associated with wage labor. In contrast, agricultural time is closely tied to cyclical natural rhythms and associated with a task orientation toward work. In Anamat, as is the case throughout much of the world, agricultural cycles are influenced by both natural and commercial pressures (see O'Malley 1990, 1992). In recent years, theoretical discussions challenging the dichotomy of agricultural and industrial temporal rhythms have increased. These criticisms suggest an alternative model of multiple temporalities contingent on contexts and activities (Adam 1995; Greenhouse 1995; Rutz 1992; Whipp 1987). Even so, there are few ethnographic or historical studies implementing such models and providing representational examples to follow (Mark M. Smith 1997 is an exception).

Anamat's diversity resists a homogenous representation of time there. The consolidation of Anamat's potential diversity into two ethnic categories and the association of these categories with temporal stereotypes force the recognition of ideological forces that associate particular ethnic groups with particular temporal attitudes. Consequently, as is the case with family organization and kinship in the Caribbean (R. T. Smith 1996:5), one is not able to attribute temporal attitudes to ethnic origins, only to the processes of maintaining and constructing ethnic differences.

### Telling Time

Trinidadians use several idioms concerning time: "any time is Trinidad time," "jus' now," and "long time." Learning the cultural ideas associated with these expressions plays a pivotal part of understanding multiple times in Trinidad. During the course of fieldwork, I conducted interviews that I began by asking informants to explain to me, an American, what Trinidadians mean when they use these phrases. I then formed subsequent questions by asking informants to elaborate on what they had said. I also conducted case studies on situations in which these phrases were used. I discovered that the uses of these expressions do not refer to a particular idea of time but instead allow Trinidadians to manage relationships organized by different models of time.

Also reflected by Trinidadians' talk is the extent to which they share metaphors of time with other English speakers. George Lakoff and Mark Johnson (1980:7–9) provide a list of temporal metaphors found in American English, a list I presented to my informants. It turns out that they share some of the metaphors, but not all. Like American English speakers, Trinidadians use metaphors such as "spending time," "wasting time," and "saving time." Other metaphors found in American English, such as "investing time," "time management," and "budgeting time," are not found in Anamat, or at least the informants I interviewed said that they had never heard of these expressions. The expression "time is money" is found, to a limited extent, among tailors and taxi drivers, and to an even greater extent among doctors and business people. The explanation for this cultural distribution is simple. Doctors and business people have the greatest degree of contact with American and English education and businesses. Tailors and taxi drivers have greater exposure to doctors and business people. The tradespeople and farmers of Anamat have almost no consistent contact with these other people. Thus, the varied distribution of the different metaphors related to "time is money" indicates the extent to which this linkage of time and capital has penetrated Trinidadian society—and the penetration has not been complete. This difference between Trinidadians is far more subtle than ethnicity, age, gender, or class, but just as important because of its representation of cultural links between Anamat and ideas of time globally distributed by institutions of education and commerce.

At the local level, cultural models of time are also expressed in activity. Some of the most trivial applications of cultural models of time reveal the degree to which humans rely upon them to manipulate their surroundings. In some parts of the world, cooks rely on timers and clocks. Sometimes, models of time do not work well, as with the difficulties some have in cooking rice—a task remarkably simple in Trinidad, where clocks are not used to time it, but remarkably complicated and potentially disastrous in places such as the United States, where timing devices are applied to the process.[1] Ideas about seasonal changes are important to farmers. In slash-and-burn agriculture, which is practiced by some Anamatians, the window of opportunity for clearing and burning land is determined by the dry seasons. There are two such seasons, *petit careme* around September and early October, and the long dry season which begins around March and lasts until June. Most farmers clear land only during the long dry season because they feel that *petit careme* is too short, although I know one farmer who feels confident in his ability to "time" his clearing and planting during *petit careme* so that he can take advantage of crops of dasheen and cassava when other farmers have none to sell. With regard to other crops, such as bananas, tomatoes, and peppers, the natural time pressures to pick and market the produce before it rots is as profound as any industrial work schedule dictated by the clock.

Just as people employ conceptions of time to cook and to work, they employ them in their social lives. The coordination of social relationships is a fundamental part of all social life, and social coordination must, minimally, involve the dimensions of time and space (Moore 1963a, 1963b; Sorokin 1964; Sorokin and Merton 1937; Zerubavel 1979, 1981). People make appointments to ensure meeting at the same time and place. Trinidadians resort to clocks to make appointments, but much of their social lives involves other means of coordinating their activities. Trinidadians employ knowledge of daily cycles of activity as often as they use clocks. For instance, the young men of the village appear at the soccer field to play at around the same time each evening. That time is approximately when they expect enough of their friends to be done with work in order to field two teams. In another example, shop owners make sure they are open as soon as they see the first schoolchildren walking down the road. This time can vary, since an inadequate water supply sometimes forces an early closing of the school. Such activities are not determined by the clock, but they still require coordination of the lives of two or more individuals.

Yet senses of time are not simply applied to activities; they also emerge out of activities. Eviatar Zerubavel (1979) describes time in hospitals in terms of the different rhythms of the patients', doctors', and nurses' activities. Representations of peasants' time suggest rhythms that emerge directly out of agrarian cycles (Bourdieu 1963, 1979; Gell 1992:86–89; Glasser 1972; Leclerq 1975; Thompson 1967). Nancy Munn argues that the Kula exchange in Melanesia defines its own time (1983:280). What becomes apparent is that every rhythmic or cyclic activity generates its own models of time. It is also often the case that people do not conceptualize or discuss these times explicitly, but instead they awkwardly relate the cycles to agreed-upon temporal standards, such as clocks. Waiting for a doctor despite having an appointment demonstrates the awkwardness of coordinating a temporal standard, in this case the clock, and a time embedded in activity. Doctors often unsuccessfully strive to diagnose and treat patients in appointments scheduled at fifteen-minute intervals. The uniformity of their schedule belies the variation in the time demands of each patient. While some societies, such as the United States, commonly relate activity to clocks, in rural Trinidad it is not as common, as demonstrated by the example of the shopkeeper.

### Time, Labor, and Production

Since every occupation has its own cycles and, thus, tacit model of time, every relationship between supervisor and laborer involves concerns about the sequence, timing, and value of work tasks, particularly when management and workers dispute the ideas of the relationship of time and work. Controlling time is controlling production. In cases of wage labor, controlling time also involves controlling wages.

The routines that emerge out of daily activities, such as work, are important for the day-to-day manifestation of Anamatian, indeed Caribbean, social organization. As Michel-Rolph Trouillot comments, "the routine repetition of certain practices leads to the formalization of relations that the actors experience as steady. That apparent steadiness, far from being a veil, is an indication of the importance of the processes that sustain those relations" (1988:231). By implication, manipulating routines can change social organization. Thus, time is not only a means of establishing routine and conceptualizing the passing of events, but it also becomes a means of sustaining social relations, or modifying them.

Where very different systems of production coexist, these relationships are further complicated. Anamat's past involves a variety of competing systems of production: slash-and-burn horticulture, plantation agriculture, small-scale cash crop farming, and industrial manufacturing to name a few. It also involves occupational multiplicity, a strategy used to combine several jobs that are insufficient by themselves to provide a desired wage level, but that in combination are sufficient (Comitas 1973).

Anamat was settled by Amerindians who practiced a form of shifting agriculture involving *conucos*—this practice is still common. In this system, farmers plant tubers. Such crops take few nutrients from the soil. The crops are left in the ground until needed, protecting them from damage due to the high winds of tropical storms. In addition, farmers raise small mounds around each plant. *Conuco* horticulture reduces erosion and eases harvesting. As one Anamatian farmer pointed out to me, if the root is in the mound, there is no need to dig a hole to harvest it; one simply inserts one's cutlass (machete) at the bottom of the mound, and "pops" the root out. In the original system, the Amerindians moved their gardens every few years and returned to an old garden site only after a minimum of twenty years. They favored returning to old plots because they found the forest there easier to clear than virgin forest (Newson 1976:39–47; Sauer 1966:51–55; Watts 1987:53–60). Now, sometimes, land is overutilized, but this rarely happens in Anamat where there is still abundant forest land—even though squatting on forest reserve land is illegal.

As has been the case with many other Caribbean islands, for much of Trinidad's colonial and postcolonial history, sugar plantations have played a major role in the island's economy. The Caribbean sugar plantation system emphasized temporal discipline and clock time. In Anamat, cocoa estates borrowed these plantation-generated ideas. The labor force, from the overseer down to the field laborers, was organized around the clock. Yet, this organization did not depend on all the members of the workforce watching the clock. Instead, the overseer kept time. As was the case on slave plantations in the southern United States (Mark Smith 1997:16) and Jamaica (Higman 1984:187), plantation field workers in Trinidad did not have timekeeping devices. Plantations signaled when to start work, take breaks, and end work by using horns or bells. This control over timekeeping devices gave overseers "the power to set the actions of others . . . against artificial, mechanical time" (M. Smith 1997:14). According to B. W. Higman, plantation timekeeping "remained in the hands of the planters throughout the period. The ringing of bells and blowing of shells that regulated the periods of work and rest was always controlled by the masters, and slaves had few means of checking the time other than the position of the sun" (1984:188). There is an account from Jamaica of an overseer destroying an hourglass used by apprentices. This led Higman to conclude "This lack of time-keeping devices provided the planters with an ultimate weapon in exploiting the slaves' labor time" (1984:189). Indeed, this sort of temporal discipline was initially introduced into Anamat by the cocoa plantations and was applied most stringently to immigrant laborers, both from India and from other parts of the Caribbean. The plantations could not as easily control the local small farmers in this way, however.

On the plantations, as in the factories of the Industrial Revolution, the multitudes of workers had to be coordinated. In factories, workers received training in a view of time as a commodity to be conserved, as well as a view that the institution, and not the individual, owned time (Thompson 1967). On the plantations, even after emancipation, a similar conception of the institution's ownership of the workers' time was promoted. In fact, the temporal discipline associated with factories in the Industrial Revolution (Brody 1989; Gutman 1976; Hareven 1982; Mumford 1963; Thompson 1967; Thrift 1981, 1988) may have arisen on the plantations of the Caribbean before the factories of Europe: in describing Caribbean sugar plantations, Sidney W. Mintz suggested, "The specialization by skill and jobs, and the division of labor by age, gender, and condition into crews, shifts, and 'gangs,' together with the stress upon punctuality and discipline, are features associated more with industry than with agriculture—at least in the sixteenth century" (1985b:47). The development of such temporal discipline in the colonial Caribbean leads to the conclusion that the factories of early industrial capitalism were not the sole institutions in which clock-driven temporal discipline emerged, and that the management of free wage labor was not a necessary condition for such discipline. As Mark M. Smith has demonstrated, mercantile scheduling pressures influenced slave plantations' adoption of the clock as much as they influenced factories' temporal discipline (1996). In effect, whether the labor force was free or enslaved was not a determining factor in whether clock-driven temporal discipline was adopted by management.

With regard to the Industrial Revolution in Europe and North America, several scholars (Brody 1989; Gutman 1976; Mumford 1963; O'Malley 1990; Thompson 1967; Thrift 1981) depict a growing emphasis on punctuality as defined by clocks and temporal discipline. In these accounts, the agriculturists who were becoming proletariats resisted industrial temporal discipline but eventually succumbed. The agricultural laborers in the West Indies followed a different evolution. As Mintz's work on the emergence of peasantries in the Caribbean shows, peasantries did not precede the plantation system in most parts of the region but emerged out of, and in contrast to, the plantation system (1961, 1973, 1974, 1979, 1983, 1985a). Plantation laborers, both enslaved and free, were coerced into adhering to temporal discipline, particularly during harvest time. After emancipation, many of these farmers attempted to establish some independence from plantations, and they grew to value self-sufficiency, independence, and their own conception of freedom. Included in this are ideas about control over one's own time. As a result, in Trinidad the emphasis on clock time and temporal discipline best applies to those contexts most influenced by global economic ties and their accompanying agents—schools and industry, for example. In Trinidad, factories commonly employ clock time to enforce work discipline (Yelvington 1995:200–202). The school system, which had its origins in missionary efforts (Campbell 1992), does so, as well. But the small-scale farming found in Anamat and the routines of many artisans and tradespeople do not emphasize clocks. Indeed, some Anamatians have remained in farming in order to be their "own boss" rather than "being told what to do and when to do it."

The coordination of exchange is as important as the coordination of production. During the Industrial Revolution in Europe and North America, the different industrial institutions had to be coordinated with one another; with the advent of more rapid transportation and the possibility of immediate communication, it became necessary to know exactly what time it was in any place on the globe. As a result, time zones were devised and put into place, to avoid the confusion created by local times which were gauged by the sun (O'Malley 1990:99–144; Whitrow 1988: 165–66). In the Caribbean, the plantations relied on merchant shipping in order to obtain needed supplies and to sell their crops. Since the plantation system was export-oriented, there was a pressing need to coordinate getting the cash crops to the docks when the ships arrived, particularly since the arrival of ships was highly seasonal—the backers and captains of sailing ships tended to avoid sailing during the hurricane season, if they could (Pares 1956:18). Even today, in parts of the Caribbean, the timing remains critical in selling produce. As Trouillot says about the sale of bananas in Dominica, "Timing is essential because of the way in which labor and transportation requirements, hard to meet on their own terms, must fit within the DBGA procedures and, ultimately, within Geest's schedule" (1988:217–18)—with the DBGA being the Dominican Banana Growers' Association, the organization that acts as the mediator between the growers and the buyers, and Geest being the fruit company that buys the majority of the bananas.

The plantations in Anamat disappeared before World War II in the wake of crop disease, falling cocoa prices, and rising wages. As a result, most residents of Anamat became farmers, either as small landowners or as squatters. The government and factories became major employers only in the 1950s, but like the plantations before, even though they both tried to enforce clock-defined temporal discipline, they were not entirely successful. Indeed, in Anamat, particularly in the Forestry Department, government work increasingly compromised on the length of workdays to permit local workers to both farm and work for the government.

Anamat, then, is neither strictly industrial nor strictly agricultural but involves a form of occupational multiplicity that includes agricultural pursuits with other forms of work. The result is that workers face conflicting time demands of production and marketing, as well as conflicting definitions of the relationship between time and work. For instance, a "day's work" for the Ministry of Public Works is from seven in the morning until three in the afternoon, but a "day's work" in the cocoa walks is from seven in the morning until around noon. When I asked why this was so, I was told, "because the work is different."

### Time and Ethnic Antagonism

In Trinidad, ideas of time reinforce ethnic antagonisms, particularly among Europeans, Indians, and Creoles. Each of these groups is associated with temporal stereotypes. Trinidadians use these stereotypes to explain behavior and to reinforce ethnic differences. Often the stereotypes cause conflict. Their origins are the labor relations in the plantation system: the relationship of time, work, and wages to ethnic differences in this system brought about stereotypes that represented different ethnic groups' attitudes toward time.

After emancipation, during the 1840s, Trinidadian planters were fearful of losing the inexpensive labor force for sugar cultivation (Burnley 1842). One colonial response to this fear was the initiation of a system using indentured laborers from India (see Laurence 1994; Look Lai 1993). Decades later, with the emergence of cocoa plantations, there arose a demand for labor in the interior of the island. This demand could not be satisfied by the squatters and small farmers of the interior who, while willing to work on plantations for wages, were primarily devoted to their own land. During critical times for the cocoa plantations, such as harvest, the small farmers and squatters tended their own land before offering their labor to planters. Squatters, especially, posed a serious challenge to the government's land policies and the plantations' ability to control both land and labor (see Chapman 1964:68–75). At the same time, planters could not request Indian indentured laborers for plantations in this region, because the plantations were on Crown lands—land owned by the government, not by private individuals. With the growing importance of cocoa as an export, the colonial administration began selling Crown lands, and once planters had legal title to their land, they could request indentured laborers. The plantations also hired immigrant workers from other parts of the Caribbean. By using immigrant labor, the planters were seeking to drive labor costs even lower (Johnson 1971:70; Laurence 1971:79; Look Lai 1993:165), and this created divisions between Creole wage laborers and Indian indentured laborers (Yelvington 1993a:10).

In addition to the antagonism generated by labor competition, on the sugar plantations Indians took jobs that the Creoles viewed as inferior (Brereton 1974:19; Wood 1968:136). The planters perpetuated these antagonisms and rankings in dealings with members of each ethnic group. Because of planters' social connections and influence, their ideas spread throughout the island, including remote places such as Anamat. These ideas were adopted in Anamat despite the fact that the predominant crop was cocoa, and cocoa's labor needs differ from those of sugar cane. These differences meant that while occupational segregation was typical in much of Trinidad (Brereton 1974:25; 1993:52), this was not the case in Anamat. There, the plantations needed to supplement their indentured workforce with Creole workers. These Creoles performed the same work as Indians and often worked with Indians. After indentured contracts expired, many Indians settled nearby, living alongside Creoles. Despite the conditions on cocoa plantations being very different from those on sugar plantations in Trinidad, the stereotypes that emerged on the sugar plantations were adopted in Anamat.

These stereotypes included ideas that Indians and Creoles possess different temporal orientations. These ideas persist even in studies of ethnic differences. For instance, in the late 1950s, Walter Mischel, a psychologist, conducted several studies on the relationship of "future-orientation" to the delay of gratification. In his studies, he compared Indians and Creoles based on their ethnic stereotypes (1958, 1961a, 1961b, 1961c). He tested the delay of gratification in Creole and Indian schoolchildren to see which group would delay gratification more often than the other. Based on the stereotypes, he predicted that Indians would delay gratification more than

Creoles. His experiments involved giving students a choice between a small piece of candy right away, or a large piece of candy to be delivered the next day. He found that Indians tended to be willing to wait a day to receive more candy, whereas Creoles tended to want candy right away. In trying to explain this difference, he turned to family structure and argued that the Indian and Creole families differed primarily in the fact that Creole fathers tended to be absent more than Indian fathers. Mischel then tested this possible correlation by comparing Trinidadian Creoles with Grenadian Creoles, whose fathers were less frequently absent, but this study proved inconclusive (1961c).

Morris Freilich, an anthropologist who studied Anamat in 1957–58,[2] noticed similar patterns. He, too, suggested that Creoles were present-oriented and that Indians were future-oriented. The evidence Freilich presented was based on the tendency among Creoles to spend more money on clothes and going to parties, and the tendency of Indians to view earnings as helping the family improve economically (1960a:145–47). This latter tendency was exhibited by Indians buying more land than Creoles (1960b).

Daniel Miller (1994) developed these ideas in the context of how Trinidadians structure the conceptual relationship of the present to the past and the future, and how these relationships emerge out of modernity. In doing so, he transformed "present-orientation" into "transience," and "future-orientation" into "transcendence." For Miller, transcendence involves developing a continuity among the past, present, and future. Examples he provides are the continuity of family, land tenure, inheritance, and the mass consumption of durable goods, such as furniture. Transience emphasizes the present, and, in Miller's analysis, an emphasis on values which are fleeting, such as short-term sexual relationships and style. In turn, he argued that the dualism of transcendence and transience permeates all of Trinidadian society, encompassing both Creoles and Indians. Moreover, Miller suggested that the stereotypes of familial, sexual, and economic behavior associated with ethnic stereotypes are linked to the dualism of transcendence and transience, with "transience" applying to stereotypes of Creoles, and "transcendence" applying to stereotypes of Indians.

These long-established stereotypes have implications for expectations concerning time. Trinidadians expect members of the different groups to live up to the group's stereotype, even though many do not. There are Creoles who are future-oriented and Indians who are present-oriented. Yet, these stereotypes are used to make behavioral distinctions, adding to commonly employed cultural classifications based on appearance. Consequently, ideas of time become more than a means of organizing experience; they play a role in conceptualizing and essentializing ethnic differences in Trinidad.

The explorations of these temporally based stereotypes fail to explore their historical contingency, however. These stereotypes emerged in a context of power relations among Creoles, Indians, and Europeans.[3] The latter group, itself stereotyped, includes plantation owners, colonial officials, and a wide variety of expatriates. The tendency to leave these Europeans out of the discussion of stereotypes suggests an ideologically driven oversight of the fact that these stereotypes emerged as a result of relationships among these Europeans, Creoles, and Indians. Trinidadians, particularly those who worked for Americans or Europeans during the colonial era, had clear stereotypes of Europeans as future-oriented or, in Miller's terminology, transcendent. For instance, drivers' most frequent comment about the Churchill-Roosevelt Highway, a highway constructed by the American military during World War II, is that it was "built to last." This leads to a further generalization that Americans build things to last. Added to these stereotypes must be the one held of Europeans as punctual and fixated on clock time. This view emerged out of European management practices on the plantations, in colonial civil service, in factories, and on the military bases during World War II.

It is also important to remember that the stereotypes emerged under conditions of European control and in the context of plantation labor. As Mark Smith remarked, slavery involved the combined use of the whip and the clock (1997:141). The reaction of enslaved Africans seems to have been resentment of such temporal discipline and efforts to undermine planters' control over time and work by resorting to work slowdowns and disruptions of the planters' schedules. During indenture, the whip was replaced by criminal prosecution against indentured laborers who did not perform at expected levels. Non-indentured workers were subject to firing or wage reductions.

The stereotype of Creoles emerged independently of the stereotype of Indians, as is indicated by its prevalence in areas with no indentured

Indians, such as the United States (Genovese 1976:295–309), but the stereotype of Indians emerged in comparison to that of Creoles. The stereotype of Creoles as present-oriented developed under conditions of plantation slavery throughout the New World, before the rapid rise of plantations in Trinidad during the last part of the eighteenth century. These stereotypes are, in large part, attributed to the antagonistic labor relations between those enslaved and their masters, and the frequent exploitative drive of planters to increase productivity, even at the cost of increasing slave mortality. The antagonism between European planters and Creole workers remained after emancipation.

In Trinidad during the second half of the nineteenth century, arriving Indians were under indenture contracts, and consequently they were under the direct control of the plantation management, whereas the Creoles (who were no longer enslaved) had established themselves as at least nominally independent of the plantations. The planters could control the indentured Indians but had to negotiate labor agreements with Creole workers, who would not show up to work if they had a better opportunity elsewhere. Indeed, Indians were contractually bound to work five days per week, and six days per week during harvest, but Creoles typically worked three days per week (Laurence 1994:306). Planters viewed the Indians as being frugal in their lifestyle as they saved for the period after their indenture contract expired. Indeed, after contracts ended, many Indians bought land and entered into arrangements with local plantations much like the Creoles, as wage laborers.

The historical conditions of the stereotyped contrast between Indians and Creoles, then, involved a context in which planters viewed Creoles as outside their direct control and, therefore, as unreliable workers, while they viewed Indians as under their direct control and, therefore, as more reliable workers than Creoles. This contrast based on the dependability of showing up to work forms one component of the temporal stereotypes of Indians and Creoles.

The stereotype of present- versus future-orientation is based historically on differing patterns of Creole and Indian consumption. The stereotype labels Creoles as present-oriented because of their expenditures on clothes and parties—neither of which is a durable good. On the other hand, the stereotype labels Indians as future-oriented because they hoarded money, which they often melted down and turned into status-enhancing heirloom jewelry (Brereton 1974:21).The Creole pattern of expenditure seems to imitate European planter patterns of status-enhancing banquets and balls (Dirks 1987:44–46). Consequently, a possible explanation of this component of the stereotypes is that both Creoles' and Indians' consumption patterns were meant to display rank, but because the objects of consumption differed, the planters' ideologically driven interpretations of the patterns resulted in different stereotypes. In the case of Creoles, these interpretations were inheritances of the system of slavery.

While historical conditions explain the origin of the stereotypes, these conditions do not explain their persistence. That is attributable to the ongoing usefulness of the stereotypes as a means of manipulating and interpreting social relationships. After the indentureship period, the relationship the European elites had with Indians and Creoles continued to differ in many parts of Trinidad, because Indians and Creoles filled different occupational niches (Brereton 1979). In Anamat, there was not the same degree of differentiation as in other parts of the island, yet the stereotypes continued to be strategically employed there as a component of power relations between planter and laborer.

### Time and Power

Planters faced the challenge of coordinating labor relations, a specific version of the more general issue of coordinating social relations. In many cases, to coordinate social relationships involves some form of agreement about the use of a single time system. Such consensus extends far beyond the organization of social relationships in time; it involves issues of power, solidarity, ideology, and legitimacy. For the most part, people agree either because they think it is right to do so, they want to do so, they feel they should do so, or they are forced. These options are not mutually exclusive. While such motivations are easily separable conceptually, in day-to-day life they are often difficult to isolate, except when force is employed, as in the use of coercion to enforce temporal discipline in the workplace. In many other contexts, people express the legitimacy of consensus, although even such expressions can, conceivably, be forced. An example of this is when workers acknowledge the importance of arriving at the job site "on time" but recognize punitive measures taken against those who are tardy.

Manipulating the relationship of time and activity is an important means of exercising power (Foucault 1977; Greenhouse 1996; Rutz 1992; Rutz and Balkan 1992; Verdery 1992). For instance, under the Ceausescu regime in Romania, the state sought to control the time of the Romanian people. Previous to the state's efforts, individual farmers had control over when to plant based on their judgment of the optimal time to do so, but as the state tried to gain control over time, planting became determined not by the farmers' judgments, but by when the state provided the fuel allotment for the tractors. Through control over electricity and water, the state controlled the scheduling of such activities as watching television, bathing, and flushing toilets. As Katherine Verdery points out, such control over time "destroyed all possibility for lower-level initiative and planning" (1992:47). This lack of initiative extended to efforts at rebellion and resistance (1992:55).

Attempts to define time are grabs at power and are often recognized as such. In contexts such as the plantation, the ability to control time was an important aspect of the planters' power, and the ability to slow down or to disrupt planters' schedules was an important element of subverting the planters' exercise of power.

In Anamat, plantation control was short-lived. Outside of such institutions, there was and is nobody with sufficient power to control the time of others. The process of coordinating social activities involves negotiation. In a place such as Anamat, with multiple work rhythms and different ethnic groups, the definition of time and the coordination of social relationships have many possible social implications for class, ethnic, and gender relations.

Aside from the general goal of coordinating one's actions with others, there are two other possible goals that motivate the manipulation of time: encouraging cohesion or encouraging conflict. So far, I have outlined several sources of division within Anamat: the varieties of ethnic divisions and occupational divisions. With regard to time, on the one hand there are the challenges of bringing people together at the same time in the same place, and on the other hand there are stereotypes about the temporal attitudes of different groups within Trinidad. As I shall show in later chapters, Trinidadians employ these possibilities in a variety of ways. For now, it is sufficient to schematically outline some of those possibilities more commonly employed.

A relationship may take place in a context where one individual has the power to control time, such as in the workplace, or in an institutionalized context where there are widely held ideas about time, such as religious services where the institution has a traditionally accepted start time. In these cases, acquiescence to the definition of time reflects acquiescence to the power of the person or institution involved. For instance, the holding of a pujah (a Hindu ritual) culminating in the lighting of deyas (small clay lamps) at six o'clock in the evening on Divali displays the recognition of the power of the sacred and the need to be punctual in meeting the demands of the ritual. Showing up punctually for a day's work during the cocoa harvest demonstrates recognition of the importance of obeying the landowner.

The possibility for the manipulation of time also provides an excellent context for publicly recognizable challenges to power. For instance, repeated tardiness to work challenges the boss's power. Indeed, showing up late in some contexts is an important component in demonstrating the lack of power of the organizers over the participants.

Another possibility involves uses of the clock to coordinate schedules and rhythms derived from several activities. In this case, those involved emphasize the importance of the relationship by scheduling it and being punctual according to an agreed-upon external measure like the clock. As in the first case, disregard for the scheduling challenges the power, and often the prestige, of the others involved.

With regard to these possibilities, the reactions of those involved are very important. Tardiness can result in open confrontation and conflict, or it can result in its being downplayed by one or both parties. In the latter situation, Trinidadian phrases such as "jus' now" and "any time is Trinidad time" play an important role. They obscure punctuality and scheduling and instead appeal to an ideological claim, albeit still highly contested, of a common Trinidadian character trait.

In other situations, such as liming, participants explicitly reject clock time. In such situations, those involved show the importance of their relationships by both parties appearing when desired or necessary, and being willing to wait for others. This situation is common in "getting organized" to attend a Carnival fete. Since the fetes are commonly held several hours' drive from Anamat, and since most fete-goers must rely on somebody else to transport them, arrangements must be made with vehicle owners. These arrangements often do not have an agreed-upon time for going but instead involve faith that the driver will show up.

## Time and Social Disruption

Ideas of time can be contested, and they can be manipulated. They also involve unintended consequences. Even when and where people seem blindly to follow a shared definition of time, they do not always do so competently. Social blunders occur, deviations exist, and such events force social actors to contend with these problems. Often, in such instances, the breakdown of coordination of social relationships in time leads to discussion and criticism.

The human tendency toward sloppy choreographing serves in the maintenance of social organization. Erving Goffman (1959, 1967) recognized that because people fear their social performances will not work, and because they take extra care in avoiding or correcting mistakes, their system of social relationships weathers the storm of repeated oversights, omissions, and blunders. When things go wrong, the amount of effort put into correcting them indicates the desire of those involved to keep things right. Knowing this desire—this expression of commitment—is very important to all involved in a relationship. Thus, breakdowns in the temporal coordination of social relationships provide possibilities for emphasizing the importance of relationships. Again, in these cases, the idioms "long time," "jus' now," and "any time is Trinidad time" are used to obscure mistakes and to emphasize commonality and cohesion.

Likewise, miscues and mistakes provide opportunities for expressing hostility, conflict, tension, and difference. By not attempting to avoid some disruptive incident or not attempting to correct the damage done by such incidents, an individual, an institution, or a group is able to embarrass and criticize others. In some cases, disruption is associated with recognized social divisions, such as ethnic or class differences. In these cases, the ethnic stereotypes play an important role in making accusations. In addition to Creole stereotypes being used to attack Creoles and Indian stereotypes being used to attack Indians, Creole stereotypes can be used to insult Indians by means of the accusation that one is "acting like a Creole." Indian stereotypes can be used in a similar fashion, although when this happens, it is also combined with an accusation of "acting big," that is, making claims of superiority. When such things happen, incidents become expressions and performances of great social significance by demonstrating differences. In this fashion, conflicts between classes and ethnic groups are expressed and maintained in daily life. Occasionally, a so-called social blunder might be purposefully performed so as to expose differences between people.

Therefore, time, as a major means of managing social relationships, serves as a means of expressing differences and similarities. If circumstances involve ethnic, class, or gender differences, stereotypes with temporal components can be effectively deployed to enhance apparent differences. Membership in any group can be performed through the meeting of temporal expectations of the group, with those who do not meet them clearly revealing themselves as peripheral members, or even nonmembers.

## Time and Social Organization

In summary, multiple times can play important roles in the coordination of a wide variety of social relationships and also in the definition of these relationships, the definition of groups, and demonstrations of solidarity, power, and conflict within a social system.

Due to processes of temporal manipulation, the study of cultural representations of time is a study of social organization. Raymond Firth described social organization as "the processes of ordering of actions and of relations in reference to given social ends, in terms of adjustments resulting from the exercise of choices by members of the society" (1964:45). In many cases, the "processes of ordering of actions" involves the attempt to coordinate social activities in time. Granted, this is a functionalist assumption, but sometimes things are the way they are because people assume they work. Trinidadians make appointments for functional reasons. They plant at particular times of the year for functional reasons. Some activities require coordinating the activities of several people, while others do not. Many Trinidadians are involved in multiple forms of production, and often these different forms of production have conflicting demands—in a sense, requiring the individual to be in two places at once. Consequently, the coordination of these activities is economically important. Yet, at the same time, because of the human capacity for sloppy choreographing, these activities are often poorly coordinated, and such poor coordination generates problems, anxiety, and frustration for many. Much of this ethnography addresses the issue of poor or absent coordination and its social significance. In a sense, while the intentions behind manipulating time emphasize function, the realities of conflict, mistakes, and unanticipated consequences form a constant challenge to function.

My point is that all societies face the problem of individuals and groups coordinating the tasks they want to perform and the tasks they need to perform. Furthermore, this is an important dimension of social organization in all societies. In Trinidad, and my guess is in many other societies as well, such coordination is very flexible, even when there are ideologies of tight scheduling. This flexibility allows a great deal of room for manipulation, and such manipulation can be used either to foster or to soothe differences.

Consequently, while the assumption that social activities are coordinated in time sounds functionalist, it actually leads to an examination of the strain which efforts at social coordination place on social relationships. Indeed, the ways of dealing with such strain become the norm, and, in cases of social conflict, these strategies often do not function either to meet needs or to maintain the social structure. In fact, the frequent use by Trinidadians of idioms such as "jus' now" and "any time is Trinidad time" to manage multiple times suggests that the strain of coordinating relationships is common.

In Trinidad, the negotiation and manipulation of time takes place in a society which manifests local and global influences, multiple ethnic groups, clear and powerful gender differences, and profound generational conflicts. Since the process of temporal manipulation encourages either cohesion or conflict, it either enhances or modifies the social differences found in Trinidad. In effect, through the manipulation of time, Trinidadians manipulate social allegiances and conflicts. While the social categories of ethnicity, gender, and class have been defined in the past and have been inherited by those in the present, contemporary Trinidadians continue to adapt to and to modify their cultural and social inheritance, and the manipulation of time is one way in which this is done.

I

# The Past in the Present

At many homes in Anamat, I met "old heads" who remembered Morris Freilich, an anthropologist who studied the area in 1957–58. They asked about him, and I updated them about his life. Then, the discussion would often turn to how things had changed since "Mr. Morris" had been in Anamat. They would say, "Long time, life was hard, but people were more loving." Then they would often add, "Not like now."

What "long time" means varies from person to person. For many old heads, "long time" refers to the period when they obtained their first paying job and any time previous to that. For instance, the old heads who worked on the plantations before World War II define their "long time" as before the late 1920s. Those who worked on the American military bases during World War II refer to that period or before as "long time." Old heads who began working in the 1950s describe that decade as "long time." Even some of those who obtained jobs during the Oil Boom of the 1970s call that period "long time." As one Indian man said, "You have to listen and let them say what they have to say, and then you will know what they mean by saying 'long time.'" He then went on to describe his own ambiguous use of "long time": "When I say 'long time,' I mean something must have happened some ten, fifteen, twenty years ago."

The comparisons that people make between "long time" and the present often have a didactic character. They contain stories of work and discipline in the past. Tellers used these stories to condemn the "idleness" of present youths. Old heads proudly state that they were always punctual and worked hard. They then complain that when they try to get young adults to work, they are late. Moreover, the old heads lament that these youths often dispute the wages they receive for work. Because of the links drawn between wages and time worked, these laments use ideas of time to compare the past and the present.

Since many Anamatians participate in an informal economy of agricultural labor, there are frequent negotiations between employers and workers about jobs, schedules, and wages. Over the past seventy years, there have been changing employment opportunities and work demands. These changes often complicated negotiations over work and wages in relationship to time. Both old heads' laments about youths and discussions about past punishments enforcing temporal discipline reflect the influence of changes in the Anamatian economy on labor negotiations.

The nature of temporal conflicts evolved with the changes of forms of work in Anamat. Anamat's first settlers were Amerindian, but, after the development of the plantation system, Africans who had escaped plantation slavery came to Anamat to achieve some degree of personal autonomy. Both groups engaged in shifting horticulture. After emancipation, the formerly enslaved Africans continued to move away from the plantations (Brereton 1992:173–74; Riviere 1972:10; Sewell 1861:117), and some came to the area around Anamat. In addition, turmoil and revolution in Venezuela motivated migration from the South American mainland to Trinidad (Brereton 1979:130–31; Moodie-Kublalsingh 1994:3). The Venezuelans, who became known as "Spanish" or "Panyols," settled in the Central Range of the island and slowly expanded into the area around Anamat.

The plantation system eventually arrived in Anamat, but not until the late nineteenth century. Because of changing relations between sugar plantations and European markets and Britain's related economic policy shift to emphasizing free trade (G. Carmichael 1961:236; Curtain 1954; Lobdell 1972; Richardson 1992:60–62; Williams 1964), many Trinidadian plantations faced a crisis in paying wages and meeting other financial obligations (Blouet 1976). This affected the French-owned plantations in Trinidad the most (de Boissiere 1992:142; A. de Verteuil 1984:7). As the French planters lost their sugar estates to British creditors in the middle of the nineteenth century, these planters began to develop cocoa plantations in the Central and Northern ranges as their source of income (Trotman 1986:198).

On these cocoa plantations, the planters and their managers introduced the same forms of industrial work discipline found on the sugar plantations; gangs worked for well-defined shifts or on tasks that were expected to take an entire day. Plantation management used bells and horns to punctuate the beginning of the workday, breaks, and the ending of the workday. The laborers who lived on the plantations were particularly subject to these rules. In Anamat, these laborers consisted mostly of immigrants from other Caribbean islands, particularly Grenada, and indentured laborers from India. These sources of labor were insufficient, however, and the planters had to rely on local labor: farmers who already cultivated their own small plots of cocoa trees or *conucos* of root crops and bananas. This coexistence meant the negotiation between the rigid temporal organization of the plantation and the seasonally variable temporal organization of *conuco* work and small-scale cocoa farming. Often, these relationships involved planter paternalism. As a result, local farmers tended to work for only one or a few nearby plantations.

The small farmers and the plantations coexisted through the boom years of cocoa. This period was from the 1890s to the early 1920s, when cocoa overshadowed sugar in its importance to the Trinidadian economy. In the 1920s, falling world prices and witchbroom disease (*Marasmius perniciosus*) struck Trinidad's cocoa industry.

## Plantations

Ganraj, an Indian old head, tells a typical narrative of those born at the turn of the century. He began working on his mother's land when he was twelve years old. She had been an indentured servant brought to Trinidad from India and then widowed when Ganraj was small. At the time, much of the land in Anamat was still forested and in need of clearing. On his family's land, Ganraj cultivated cocoa and "short crops." These short crops were mostly vegetables harvested two or three months after planting, rather than in the six to nine months it takes most Caribbean root crops to mature. When Ganraj was older, he worked for an estate a couple of miles away. The overseer of the estate had befriended Ganraj's family. In addition, according to Ganraj, the white French Creole owner was a "gentleman." Whenever Ganraj needed work, the owner provided Ganraj with a job. Ganraj valued this, because he continued to work on his family's land and only worked on estates when his family needed cash. The "arrangement" Ganraj had with the owner of the estate allowed Ganraj to maintain his family's land and earn wages.

The estate owned by the French Creole man for whom Ganraj worked was about two and one half miles away from where he lived. Closer to Ganraj's home were estates owned by Spanish farmers. These farmers sold their land between 1910 and the early 1920s. Soon after, Charles Conrad Stollmeyer purchased and consolidated this land. Stollmeyer was a prominent cocoa baron in Trinidad whose home in Port of Spain, "Stollmeyer Castle," remains an icon of past planter opulence (see A. de Verteuil 1994:92–135). After Stollmeyer's company took over, Ganraj occasionally worked on that plantation, as well as on the French Creole's plantation with which he had his "arrangement."

According to Ganraj, the French Creole's plantation employed about twenty men and fifteen women from Anamat. Ganraj recalls that "there weren't enough people in [Anamat] to supply labor for all the estates." Consequently, much of the workforce consisted of immigrants brought to Anamat by the plantations. These workers were either indentured Indian laborers or immigrants from other West Indian islands. On the French Creole's plantation, there were six "barracks." The overseer lived in one of these buildings, and the other five housed Indians and workers from the nearby island of Grenada.

The Grenadian workers were not bound by any contract and could be fired. On the other hand, the indentured Indian laborers were bound to the plantation by a contract. The terms of the contract included the estate providing food, housing, medical care, and wages in return for the Indians' work. These contracts had other terms, as well. The freedoms of the Indians were restricted. For instance, they could not leave the plantation without a pass. In addition, they were subject to criminal prosecution if plantation management thought they avoided work (see Brereton 1985; Laurence 1994; Look Lai 1993; Trotman 1986). This created a situation in which work discipline for indentured laborers was maintained through recourse to criminal law; work discipline for immigrants from other West Indian islands was maintained by eviction and firings; and work discipline for Anamatian settlers was maintained by suspensions, wage reductions, firings, and the termination of "arrangements" such as Ganraj's.

The workday began at seven in the morning. An hour before, the plantation sounded a signal. Some plantations used the blowing of a conch shell for this, while other plantations used a horn, and still others used a bell. At seven, the workers had to be at the plantation in their work gangs. The overseer took roll and gave the "driver" (a term left over from slavery) of each gang his instructions for the day. Ganraj and his peers report that if a worker arrived after the seven o'clock signal, he or she did not get work that day and consequently lost one day's wages. Writers on immigrant labor state that if these laborers did not report for work, they were disciplined by a loss of pay and a record being kept (Trotman 1986:191).

In assigning work, the overseer would determine whether each gang was going to perform "day work" or "task work." Day work required that the workers toil the entire day until five in the afternoon. Plantations allotted workers a break for lunch, indicated by the sounding of the plantation's signal at the beginning of the break and at the end. It was the responsibility of the driver, a worker trusted by the planter and overseer, to ensure that the work gang labored throughout the day. Ganraj says that when he served as a driver, it was quite common for the overseer to visit his gang during the day to inspect the tasks performed and make sure that the gang members were working.

Task work involved overseers assigning specific amounts of work for the gangs to complete. Plantations paid gang members for the completion of the task. For instance, in clearing undergrowth from around trees (cutlassing and arrondeering), workers were paid per tree. For digging ditches to keep the cocoa trees well drained, workers were paid based on every one hundred feet of drains they dug. Even though planters and overseers calculated wages according to the amount of work completed, they still structured the day according to the plantation's clock. That clock determined when gangs began work and when they took breaks. Furthermore, as happened with day labor, the overseer visited the gangs to inspect their work.

The overseer attempted to maintain control over conceptions of time on the plantation. Old heads recall that he was the only one who carried a watch. One elderly Indian recalled an incident where the overseer confiscated a worker's watch. This gave workers the impression that the power to employ granted the employer the power to define time. Some members of this generation of old heads still manifested this attitude. Tantie Oakley, an elderly woman, occasionally employs men to help her with her land. One morning, one of her workmen arrived at 9:30 A.M. She greeted her worker by saying, "See how late you come? It nearly ten o'clock." She then gave him the list of tasks she expected from him that day, ending "You lucky I keep you on here. No one else would hire you or be as good to you as I. Now it is nearly eleven o'clock, so get to work!"

## The Demise of the Plantations

The period during which Ganraj began to work on plantations was a boom period for cocoa. Indeed, cocoa was Trinidad's "most important export" (Phillips-Lewis 1988:29). This ended in the late 1920s. During this period of prosperity, poor management plagued the large estates: most plantations did not build reserves to mitigate against falling prices or crop disease; they spent all their profits; and they kept poor records (Shephard 1932:307–17, 334–45, 1936:327–29; Singh 1994:82). For the most part, overseers and plantation owners were unaware of the exact amount of expenditures necessary to maintain their plantations, and they did not know the relative yields of individual fields (Gilbert 1931). As a result, they could not know how to most effectively spend money to maintain their holdings and maximize productivity. During a period of high prices and low wages, such poor management practices did not seem harmful, but when the prices fell and wages rose in the 1930s, such practices proved disastrous (Singh 1994:82–83). In the late 1920s, a devastating crop fungus known as witchbroom disease struck cocoa trees throughout Trinidad. The only effective means of treatment was to cut off affected parts of the tree. This cure required much labor. By the early 1930s, cocoa planters were in a difficult situation: they needed a large labor force to fight the disease; labor was more expensive than in earlier years; and cocoa prices were falling. To deal with these factors, many plantations borrowed money from the colonial government's Cocoa Relief Board, but conditions were such that many of these plantations never recovered financially.

During the boom years, many cocoa dealers sprang up in the village independent of the planters. These new middlemen often ran multipurpose establishments combining the purchase of cocoa with the selling of groceries, dry goods, and liquor. In this way, entrepreneurs arose from the class of small-scale farmers and began to accumulate land by issuing loans against expected harvests. Shopkeepers took advantage of the drinking habits of some of the residents of the village, as well. Loans grew not only from the sale of food, dry goods, and supplies on credit, but also from the sale of rum on credit. In a few cases, several members of the same family who drank heavily all borrowed against their harvest and their land at the same store. Soon, their holdings were deeply mortgaged. Eventually, the store owner seized their land as payment for their debts.

The Chinese and the Indians tended to be the groups that participated in this retail trade. Members of these groups accumulated land and developed the largest estates, aside from those owned by the whites. The Indian estates did not utilize indentured labor, however, and relied on kinship networks and wage labor in order to manage the land adequately.

Harban, an Indian man about fifteen years younger than Ganraj, recalls the role his father played during this time. Harban's father had a shop and made loans. He was able to acquire a large amount of land in 1933 from a French planter who had financial difficulties. Harban's father managed the estate and eventually opened a second shop about a mile and a half away, leaving the first shop to be run by his wife. Harban describes his mother as a "business woman." She would extend credit, and every night Harban would record how much credit she gave during the day and what payments had been made toward the debts she was owed. When she grew old, she sold the shop to another Indian entrepreneur who owned land and wanted to enter the retail business.

What Harban's memories suggest is that the ruin of the white planters provided an opportunity for locals who had been accumulating cash to acquire more land. Indian families eventually bought all the large estates, with one exception. The one exception was a plantation that became so encumbered with debt from the colonial government's Cocoa Relief Board that the state-sponsored Agricultural Credit Bank foreclosed in the early 1940s, and the government obtained a legal title to the land in 1949. Acting upon recommendations found in several government-sponsored reports (Agricultural Policy Committee of Trinidad and Tobago, 1943; Jolly 1954a, 1954b; Shephard 1954), the colonial administration chose to lease this land in five-acre parcels to local residents. Seymour, a Creole man, tells how his father had a contract to do work on that estate and how, when the government took it over, he did not receive any compensation for his work. He eventually obtained one of the five-acre parcels, but soon after he "went mad." Seymour says that his father went mad because he would give out loans and work for others but not receive payment, and that this caused him to "suffer too much."

Local small-scale farmers continued to borrow from local businesses and from other farmers, usually kin. This resulted in some consolidation of family holdings, but few bankruptcies. A major advantage the small farmers had over the large planters during this period was that they did not have to rely on wage labor to combat witchbroom disease. Thus, the small farmers survived the period of poor cocoa prices and disease much better than did the large plantations.

Factors such as the fall in world cocoa prices and the global economic depression of the 1930s affected the people of Anamat. These global issues caused the decline of the plantation system which meant the disappearance of a system of patronage that was Anamatians' major source of cash. The last indentured laborers' contracts were terminated in 1920. These Indians were now free to work wherever they wished, if they could find work. This was a difficult time, and many Anamatians shifted their emphasis away from the cultivation of cash crops to the cultivation of food. Those who had land concentrated on its cultivation, while most others relied on squatting to produce food for personal consumption and a little bit of local marketing. The government encouraged this by recommending that farmers cut down their cocoa trees to replace them with food crops.

With regard to the connection between work and conceptions of time, Anamatians were left to define their own time. As is the case in other parts of the Caribbean, when one source of employment is insufficient, people resort to occupational multiplicity (Comitas 1973). The old heads from this era report an opportunistic view of work—taking whatever they could get. On this basis, they condemn contemporary underemployed youths' seeming lack of desire to find work or lack of gratitude when offered jobs. There is one major difference between the 1930s and the present, however. There was a shortage of cash in Anamat in the 1930s, and this forced reciprocal exchanges of goods and services, as well as credit. The extension of the latter occasionally ended with the foreclosure on land. There is enough cash in contemporary Anamat for young adults to expect wages which are often higher than landowners are willing to pay, and for merchants to demand cash payment. Contemporary merchants extend credit to old heads but almost never to young adults.

## World War II

People like Harban, Ganraj, and Seymour's parents were committed to farming by the time of World War II and did not seek employment on the American military bases. However, many younger Anamatian men did work at Fort Read and Waller Field, the installations nearest Anamat. In fact, one elderly Spanish woman reported that only the women, the old, and the children did not work on the bases. The U.S. military supervised the construction of the bases during the period of 1940 to 1942, as a result of an agreement between the United States and Great Britain. This agreement involved the United States giving the United Kingdom fifty destroyers in return for ninety-nine-year leases on land throughout the British Empire. Implicit in the agreement was the United States' ability and willingness to patrol and protect British territory and shipping lanes, particularly in the Caribbean. In the case of Trinidad, the U.S. military defended Trinidad's oil supply and Guyana's bauxite supply, both of which were crucial to the British war effort (F. Baptiste 1988; Johnston and Shoultz 1945–47). Although the war started far from Trinidad, its global influence affected Anamatians' way of life.

The American military offered wages far above those typical for agricultural work. Even though working on the bases involved a long commute on bicycle, most young Anamatian men sought such jobs because of the lure of the high wages. While there was no doubt that the employment opportunities on the bases pulled workers away from agriculture, the cocoa plantations in Anamat were on the decline anyway, and workers were unable to rely on plantation work to obtain cash. Because of food shortages, the government's recommendation to replace cocoa with food crops gained a sense of urgency. In a sense, while the military bases struck the death blow to the plantation system in Anamat, the system was already dying before the bases began hiring.

Reynold, a Creole man, worked at measuring and cutting wood in the sawmill for Fort Read, an army base near Arima. He recalled that, when construction began, he heard rumors that there would be work for ninety-nine years (presumably because the lend-lease agreement involved a ninety-nine-year lease), and that all one needed to do to get work was go to the base's employment office. Once a potential employee arrived there, "a fellow" would choose who would go through the screening process. One of Reynold's Indian neighbors, Ramran, remembers that when he first applied for a base job, he arrived at the base at nine in the morning. He then went with the other potential hires to the employment office. They arrived there at 9:30. They were given physicals. If they passed the physical they were assigned identification numbers and had their photos taken with their numbers. This formed the basis for identification badges they were issued and told to wear on the base. Ramran remembers that, by the time the hiring process ended, and Ramran and his peers arrived at the work site, it was 3:55 in the afternoon—only five minutes before quitting time. Even so, Ramran remembers that everyone was paid for a full day's work.

Narine, an Indian man, says that one of his most vivid memories of working on the base was the "pressure" involved in getting to work on time. He remembers waking at four in the morning, in order to ride his bicycle to Cumuto to catch a truck to the base. After finishing work at five in the evening, he returned home on his bicycle.

With the time demands of getting to and from work, it became extremely difficult for men to uphold the occupational multiplicity that they had relied upon previously. They became committed to a single source of work as long as they worked on the base, and that employer possessed their time. In some cases, their work was highly structured and closely supervised, such as in the construction of the buildings. In other cases, soldiers used Trinidadian workers as smugglers and look-outs, to foster officially prohibited activities such as drinking rum and gambling. By 1942 many of the Anamatian workers were no longer working in large gangs but as individuals for different enterprises on the base. Indeed, listening to their stories, I got the impression that, as the buildings on the base were completed, the soldiers preferred breaking up the gangs of workers into small groups that could act as personal servants.

The military police searched workers entering the base, to control the illicit and profitable trade in rum, cigarettes, and food. From Anamatians' perspective, the American soldiers consumed large quantities of rum, but officially rum was not allowed on the base. The soldiers obtained rum by hiring their Trinidadian workers to smuggle it onto the base. Ramran remembers deciding that if he could smuggle cigarettes off the base, then he could smuggle rum into the base to sell for a profit. One day, on the request of his supervisors, he made nine trips. During the last trip, the military police caught him, drank his rum, and warned him not to try smuggling again. He said that he could buy a flask of rum outside the base for fifty cents, but that the soldiers would pay one dollar for the same flask.

Meanwhile, the island experienced a severe shortage of food because shipping was devoted to military cargo rather than foodstuffs (Anglo-American Caribbean Commission 1943). Many staples, such as rice, were rationed. Roger, a Creole man, commented that, once at the base, he had to "guard his lunch," or else it would have been stolen. The wartime conditions had created food shortages throughout the island. The government initiated what it called the "Grow More Food Campaign," encouraging farmers to convert from cash crops to food crops. One Indian man quipped that "in those days, breadfruit was placed in a showcase for sale," and this was despite breadfruit being viewed as one of the least palatable sources of starch available in Trinidad. In Anamat, because cocoa was not as profitable as it had been a few years earlier, some plantations allowed portions of their land to be cleared and planted with food. Typically, men would work on the base, and women would cultivate the land and grow food.

The division of labor that emerged during this period created new demands on coordinating different domestic and remunerative activities. The men's days were filled with work and their lengthy commute on bicycle to and from the base. Consequently, they were unable to work their land, much less to work for nearby plantations. Women had traditionally worked on the plantations and family land, as well as seeing to domestic tasks such as cooking and child care. In many cases during the war, women filled the voids left by men who worked on the base. This meant more work for them to do. Since the base workers had to begin their commute much earlier in the morning than they had set off to work on plantations, and since there was the expectation that women prepare their lunches, the women in a household had to rise long before sunrise to make food for the men working on the base. After cooking, these women went into the fields to grow food. This food then supplemented the small amounts each household was rationed. Men then expected women to have a meal ready soon after the men arrived home. Lydia, a young wife during World War II, remembers women awakening "at all hours of the night" to cook. The women toiled in kitchens poorly lit by kerosene lamps and cooking fires, in order to prepare lunches by four in the morning for men to take to work. In the evening, when men arrived back in Anamat, the women again cooked in the dark.

The Grow More Food Campaign succeeded (Marsden 1945:33). This was largely attributable to the efforts of women and elderly men who did not work on the bases. Their efforts also meant that the money that families accumulated during the war was not simply from the wages men earned at the base but also from the agricultural efforts of these workers. The possession of food also took on great significance for Anamatians at the time when much of the rest of the island experienced food shortages.

By 1942, opportunities at the nearby bases decreased. As several informants said, nobody was fired, but instead, the base shifted the jobs to Port of Spain and Chaguaramas. Commutes to these new locations were too long for men living in Anamat, so they stopped working for the American military.

The men who worked developed very different senses of time than those who remained in agriculture. Since the demise of the plantations, agriculture emphasized occupational multiplicity—the pursuit of several sources of money. The base workers, on the other hand, were devoted to a single source of wages, and to employers with complete control over employees' time. The workers remember their days as being devoted to getting to work, working, and the long commute home. Among these men, there were no opportunities for managing several different jobs. In their complaints about contemporary youths, these men complain about the youths' lack of dedication and discipline. For instance, Ramran complains about some of his young adult workers, "When I tell them I need them seven in the morning, I don't mean come at nine, but they don't study that—to them, any time [is] Trinidad time."

Political changes soon followed World War II. In 1946, the British Colonial Office instituted a set of political reforms in Trinidad (Ryan 1972:65–67; Singh 1994: 186–222; West India Royal Commission Report 1945). These reforms included extending the right to vote to all adult Trinidadians and the election of some of the members of the colony's Legislative Council (Blanshard 1947: 115–16). Previously, all members of this council had been appointed by the colonial government. The Legislative Council consisted of prominent Trinidadians who proposed laws to the governor and advised him on issues of importance to the colony. Under colonial rule, the Legislative Council possessed influence, because it represented the wealthy elites of the island. The election clearly moved Trinidad toward democratic self-rule, and although the governor still retained his powers, several of the elected members of the Legislative Council gained considerable sway in determining public policy.

Ethnic antagonisms between the Creoles and Indians in Trinidad, including Anamat, played a role in the 1946 elections. There was ethnically motivated political violence in the market town nearest to Anamat (La Guerre 1972). In Anamat's district, the election was a contest between Victor Bryan, a wealthy white Creole planter, and Joseph Moonan. Ramran looks back at this election bitterly. He views it as the origin of racism in Anamat. He surmises that, early in the campaign, Bryan calculated that there were more Creoles in the district than Indians, and this led Bryan to tell Creoles that they needed to "protect their own." Ramran remembers Bryan as a "terror to Indians" who later "changed sides" by joining the Indian-dominated Democratic Labor Party.

Victor Bryan won the election and quickly gained a reputation for providing small favors to the residents of his district (Ryan 1972:96; Lewis 1968:209). Winfred, a Creole man who moved to Anamat after World War II, remembers Bryan as "doing work for the district." What lured Winfred to Anamat was the expansion of the government-run teak plantation and the availability of land on which he could squat and "make a little garden."

There were other government-sponsored projects that provided employment during this period as well, such as the eradication of malaria. Dr. E. de Vertueil, a member of one of the most prominent French Creole elite families in Trinidad, and also a family with cocoa plantations around

Anamat, was a leading authority on malaria in Trinidad. By the time of his retirement in 1941, de Vertueil had shown that the *Anopheles bellator* mosquito was the primary carrier of malaria in the cocoa-growing areas of Trinidad. He was also instrumental in obtaining Rockefeller Foundation support for the study and elimination of malaria from the island. This effort began in the 1940s with the work of Dr. C. S. Pittendrigh. Pittendrigh chose Anamat as one of his field sites, and he hired Anamatians to help in his study of the malaria-carrying mosquitoes of the area (Waterman 1967).

Many who were young adults during the 1940s, but too young to have been employed on the American bases, remember "Dr. Pitt's" arrival. Ramkaran, an Indian man, remembers: "The first time he came down with a Dodge van. You know, he watch us up and down. We wasn't rude or anything like liming on the block or anything, but he used to see we playing and rolling box cart and all this. He said, 'All you come, that have no future—come and do something, get some work.'" Pittendrigh hired people to catch mosquitoes for his studies. He paid boys to climb a tree in the cemetery with a fine-mesh net in the evening. Underneath the tree, Dr. Pittendrigh placed a donkey. The boys then dropped the net on the donkey when Dr. Pittendrigh instructed them, thereby catching all the mosquitoes attracted to the animal. After his work capturing mosquitoes off donkeys, he had local men build a large platform. On this platform, he had the men put bromeliads (parasitic plants that grow on tree limbs) filled with water. Periodically, he then had the men bring him the bromeliads, and he studied the mosquito larvae that grew in the water. After determining that the malarial mosquitoes bred in these pools of water, he recommended a campaign of spraying bromeliads; this program successfully eliminated malaria in areas like Anamat (Waterman 1967) and employed many young men.

According to Ramkaran, Pittendrigh had an important effect on young men in Anamat:

> He get all the boys [he hired] after to get permanent [work]. A good bit of them still with him [with the anti-malaria division of the Ministry of Health]. He break them from idleness. They never have time to study "Trinidad time is any time." You know, [when

it comes] time you have to go four o'clock to get your twenty-five cents. O God! Beat yourself and go in the cemetery . . . and he break off we in a working gang, you know what I mean. We come out working—we never idle. When he leave, he had some boys who want to go with he. They leave from here and went [to Viego Grande]. Look, most of the boys now my age [late sixties in 1990], I would tell you, they went.

Consequently, through the anti-malaria campaign and the teak plantation, employment opportunities emerged which replaced those lost as the cocoa plantations reduced their workforces. Victor Bryan encouraged and sponsored the expansion of small farming in Anamat as well. Ganesh recalls an incident where he went to Bryan during the continuation of the Grow More Food Campaign in the 1950s. At the time, Bryan was the Minister of Agriculture. Ganesh argued that he would grow food if given some land. Bryan, who knew Ganesh and his family, worked out the details and granted Ganesh some land in the form of a lease in which Ganesh paid $40 TT per year, and $185.50 TT at the end of each twenty-five-year term of the lease. Ganesh had the option to renew, and in addition, he rented additional land at the rate of $350 TT per year.

Because of his effective use of patronage, the district continued to support Bryan. As a large landowner, Bryan occupied a transitional role in shifting the source of patronage from the planters to the government. In many ways, Bryan's patronage was a substitute for planter patronage, in that it emphasized access to land and agricultural jobs.

The wage opportunities in Anamat during this time allowed workers to return to the patterns that existed during the peak of the plantations, namely, seeking wage work to supplement their household incomes. One of the significant differences between these two periods was the price of cocoa relative to the wages offered by Dr. Pittendrigh, the teak plantation, and the expanding Department of Works that was in charge of improving the roads. During the plantation era, when cocoa prices were high, farmers worked their land, but when cocoa prices were low, they sought wage labor on the plantations (Shephard 1935:85). After World War II, cocoa prices were never high enough relative to government wages to warrant choosing working one's land over the opportunity of a government job.

At the same time, there was pressure on the local overseers and foremen to build flexibility into government work to allow workers to continue to maintain their land. In a sense, this, too, became a form of patronage.

### Patronage and Time under the People's National Movement

The period between World War II and the 1970s was one of growing employment opportunities for Anamatians. During this period, the different forms of political patronage offered to Indians and Creoles reinforced stereotypes of each group. With regard to Creoles, the jobs most open to them as a result of political patronage also allowed for temporal flexibility and occupational multiplicity. These opportunities were one of the government's attempts to gain and retain political support. The opportunities most open to Indians emphasized agriculture or were through institutions separate from the state, namely, the denominational sponsorship of public schooling, particularly the Presbyterian and Hindu schools—the only schools in Anamat. The opportunities which schools generated for both teachers and caretakers emphasized temporal discipline.

Between 1946 and 1956, the government of Trinidad was primarily in the hands of a colonial governor in consultation with the partially elected, partially appointed Legislative Council. In many cases, council members such as Bryan colluded with the governor in running the colony. In 1956, the People's National Movement (PNM) gained control over the Legislative Council. The PNM's leader, Dr. Eric Williams, demanded and obtained control over internal affairs within the colony, thereby accelerating the process of gaining independence. Under PNM control, the government increased its role as a source of employment (Ryan 1972:386).

In 1956, one Indian Anamatian, Mr. Monansingh, made a living fixing cars and cultivating his five acres of land. The local Public Works overseer would go to him to have his car fixed. One day, the overseer asked Mr. Monansingh if he wanted work, and he said yes. Soon afterward, Mr. Monansingh began to encounter problems. The overseer assigned him to paint bridges, but the mason who was in charge of building the bridges was a Creole man who did not want to work with an Indian painter.

The "racial problems" Mr. Monansingh remembers involved his Creole boss imposing burdensome temporal standards in hopes of finding some justification to fire Mr. Monansingh. Like other Anamatian workers at the time, he tried to maintain his cocoa land and desired a job that would pay good wages and provide enough flexibility for him to farm. In addition, the wages offered by government work were greater than what he could make cultivating his five acres of cocoa. As a result, he was motivated to take the government job even in the face of discrimination. He remembers the time demands well, however:

> When I started working Works Department, I used to work from seven to half past four. From there, I used to ride a bicycle to go to Biche. That is about 15 miles from Viego Grande. To reach there for seven o'clock and leave there half past four to come up, riding a bicycle, and we couldn't do otherwise, we had to do it. In those days it [the time it took to ride] was not okay, because, and especially if, someone is spiting you. Well, this fellow [the Creole mason], well, I have to say he was fighting race. Now, in those days, you as an Indian could not have get employed in the Works Department as a tradesman. You couldn't. And I happen to get there through some favor, so I have to be very, very, very careful, otherwise they would replace me with somebody else of their own choice.

He then went on to relay how his immediate boss, being Creole, eventually laid him off and replaced him with a Creole. Mr. Monansingh complained to somebody in the Ministry of Public Works, who was able to get Mr. Monansingh rehired, but from that point on, the Creole bosses "timed" Mr. Monansingh. In his view, his Creole boss hoped that Mr. Monansingh would arrive late to work or quit early, thereby providing an excuse to fire him.

There did seem to be favoritism toward Creoles by government departments in Anamat. In contrast to Mr. Monansingh, Mr. Smith, a Creole, recalls that all he had to do to get work with the Public Works Department was to go to the district headquarters and ask for work. Mr. Fernandez, a Creole with some Spanish ancestry, recalled that it was very easy for him to get work at the teak plantation, while Ramran remembers that the teak plantation hired him only when it was desperate for workers and that he lasted only one day working there.

Even though employment opportunities differed for Indians and Creoles during this period, they did agree on a ranking of occupations. According to Freilich, who studied Anamat in the late 1950s, Creoles and Indians shared values of independence, "being known by the government," and not having to work hard (Freilich 1960a: 35). The emphasis on independence, which had been a hallmark of Anamat stretching back to its obscure beginnings, remained important. The ranking of occupations Freilich reports is:

high prestige: teachers

chief overseer of government work

peasant farmers

overseer of the teak plantation

shop keepers

semi-peasants (holdings of less than 8 acres who supplemented their income with wage labor)

taxi drivers who own their own car

taxi drivers who drove another person's car

road work overseer

workers for the Malaria division

road workers

day laborers. (1960a: 36)

In this ranking system, "peasant farming" retained a great deal of prestige because it offered independence. Many of the low-prestige occupations were the direct result of government patronage; "day laborers," people who worked on plantations also held one of the lowest prestige jobs. By the 1980s, economic opportunities generated a change in values, so that agricultural occupations had low prestige, while government work, even road work, was highly valued.

In Anamat, a pattern emerged in which the Department of Public Works and the Department of Forestry both hired Creoles. Locally, the Malaria Division of the Ministry of Health tended to favor Indians. Expatriate supervisors dominated the Malaria Division and the Department of Public Works. These supervisors emphasized a structured, disciplined approach to time that was similar to the ideas of time inherent in plantation labor.

The Department of Forestry, on the other hand, had Trinidadian supervisors. The work there allowed for temporal flexibility. Indeed, many of the workers on the teak plantation were also able to cultivate their land easily.

As Freilich reports, Anamatians emphasized independence, but this emphasis became linked with ethnic stereotypes of temporal attitudes in different ways for Creoles and Indians. According to Freilich, Creoles emphasized "freedom" and a "present-orientation" in which "the present is important for its own sake and the future only becomes important when it becomes the present" (1960a: 144). Indians had a "future-orientation" manifested by a commitment to the long-term economic success of the family (1960a: 144–46). In some respects, this involved a Creole emphasis on individual accomplishment and displays of individual wealth versus an Indian emphasis on family accomplishment and displays of family wealth.

Whatever the psychological validity of these temporal differences, the political and economic realities of the 1950s continued to reinforce the ideology of such differences through different employment opportunities offered to Indians and Creoles. The patronage the PNM government offered took the form of wage labor for the Department of Public Works and the Forestry Department. The patronage offered by Victor Bryan, who remained Anamat's representative, but who now also served in the new parliament, emphasized agriculture and land holdings. The opportunities open to Creoles encouraged individual occupational multiplicity. This made individual temporal flexibility desirable with regard to any job, and the Forestry Department work and, later, the Department of Works jobs began to reflect this. The opportunities most open to Indians encouraged household occupational multiplicity. This made temporal flexibility less necessary but family commitments more functional. As a result, members of Indian households began to engage in a wide array of occupations, ranging from shop keeping to taxi driving and almost always including cocoa farming. Significantly, Indians, such as Ramran, who could not rely on extensive kinship networks for work and money, tended to prefer working for the Department of Works and other jobs with temporal flexibility.

These different occupational patterns influenced the scheduling of work. When there are multiple sources of employment available, individual occupational multiplicity forces the individual to manage competing time demands from different jobs. This often involves compromise and negotiation. On the other hand, household occupational multiplicity allows some members to hold jobs which require a strict sense of temporal discipline.

### The Temporality of Patronage

Dr. Eric Williams, the prime minister of Trinidad and Tobago from 1956 until 1981, wanted to develop community-based village councils that would serve as links between the populace and the government. When Freilich was in Anamat in 1957–58, the village council was a poorly functioning body with little budget, little influence, and few successes. The villagers distrusted anyone who sought power, thereby making anyone who sought a position on the village council suspect. Consequently, the council received little support (Freilich 1960a:99). This changed, however, when the PNM-led government's approach to building the infrastructure of the island started to emphasize local government in the early 1960s, through the "Better Village" program (Craig 1985).

In the government's efforts at economic development, it made improvements in the infrastructure of the island a priority. It was felt that improvements in the transportation system, electrical service, telephone system, water system, and roads would make Trinidad and Tobago attractive to foreign investment (Carrington 1971:135; Ryan 1972:384–85). As the PNM's program gained momentum, Anamat's village council's role as a broker for government benefits grew. The council requested pipe-borne water, paved roads, electrical service, and a community center. The government granted all the requests. The village council also demanded that Anamatians be hired for projects in Anamat. The government acquiesced and gave the council a role in hiring workers. For a short period of time, the village council became a locally powerful group, albeit not without the distrust of those seeking power that had plagued it before.

Anamat's infrastructure changed considerably during this period. With the improvement of the road, and the accessibility to pipe-borne water and electricity along the road, many households moved from their own plots of land to land that they rented along the road, so they could take advantage of the new amenities found there. Many Anamatians obtained government work. Almost every household counted on at least five or ten days of temporary work from the government every Christmas and Carnival season.

During the plantation era, plantation work served as a means for small farmers to obtain cash through wage labor but, by the 1960s, government work filled this function. Most Anamatians took advantage of temporary work projects, but some, particularly young adult men, sought permanent positions with the government. All the while, Anamatians continued to make cultivating their land their top priority but supplemented their agricultural income with government wages.

As a result, almost every household had to manage conflicting time demands of government work that expected full days of work Monday through Friday with agricultural work that was intense during some periods and during other periods not demanding at all. At this time, government work evolved toward greater temporal flexibility through shorter work hours.

## Oil Boom

Trinidad's economic situation changed considerably in the early 1970s. During the world oil crisis of that time, Trinidad, as an oil-producing nation, reaped huge financial benefits. Fueled by petroleum revenues, the PNM government shifted its development policy from emphasizing the invitation of transnational corporations to build plants in Trinidad to emphasizing massive state participation in the economy (Premdas 1993:101; Sutton 1984:49–52, 60–66).

"Money flowed like water," Anamatians recall. Trinidadians purchased luxury items and even traveled to the United States to do their shopping (Miller 1994:204–5). Most of the money spent on consumer goods was spent on imported goods. Before the Oil Boom, V. S. Naipaul observed, "To be modern is to ignore local products and to use those advertised in American magazines" (1962:46); he added, dryly, that Trinidadians who grow coffee drink Nescafé instant coffee. Cash generated by Oil Boom employment increased the numbers of Trinidadian consumers pursuing modernity through consumption. During the boom, a columnist summarizing the views of Dr. Eric Williams on this matter wrote, "People begin to live and think in metropolitan ways, they aspire to the 'modern' lifestyle that they see in cinema and TV advertisements and on their increasingly frequent holidays in New York and Miami, and they try to buy the same life for themselves at home" (Frankson 1981 : 112–13). Even so, the government remained unconcerned about rampant consumer spending on foreign goods.

As was the case with the previous periods, the Oil Boom left its mark by creating employment opportunities distinctive to that period. The generation reaching adulthood during this time also differed in other ways from its predecessors and successors. The PNM government promoted education, and the opportunities to attend school beyond primary school expanded considerably in the 1960s and early 1970s. For Anamatians, the nearest secondary schools were located in Viego Grande. Since the road to Viego Grande was in poor condition, and transportation to Viego Grande unreliable, secondary school students did not live in Anamat and commute to school. Instead, they lived with family or friends who resided close to their school. The achievement of education at the expense of help in the family's agricultural pursuits became a consequence of this pattern. Many of the students who left Anamat for school left Anamat for good and sought work in the nonagricultural sectors of the economy.

With the burgeoning economy of the mid-1970s, considerable opportunities in construction and manufacturing emerged for those who had just completed their schooling. During this period, Samuel, a Creole man, obtained work at the Amalgamated factory outside Arima, not too far from the sites of the American World War II bases of Fort Read and Waller Field. The Amalgamated factory assembled cars, particularly Fords. Samuel quickly rose to the position of foreman. With government money pumping into the plant, job opportunities increased, and Samuel recruited his friends and relatives from Anamat to work at the plant.

Arima was a long commute for these men, and particularly difficult for those who had to work the 6 A.M. shift. Simraj, an enterprising Indian man who had started working a taxi between Viego Grande and Anamat after World War II, took advantage of this opportunity and bought a bus to carry the men to and from the plant. This was a valuable service to these men, and it became an important form of patronage that ensured Simraj's repeated election to the County Council. From his position on this council, he convinced the government to spend additional money on Anamat, improving roads and building a recreation ground with covered, concrete stands. All of this created additional employment opportunities for those not working at Amalgamated.

The shift workers at Amalgamated used a time clock. This differed from the system employed by plantations and government departments. Their system involved an exchequer who signed in each worker at the beginning of the workday. Under this system, being late for work often meant not being able to work that day. With the time clock at Amalgamated, however, workers could be late, and "the boss would cut your time." Frequent tardiness would result in a warning letter from the supervisor. Occasional tardiness might be covered by another worker: "Your next partner who did not reach late will clock your card for you. Like my partner reach seven o'clock, he clock my time card seven for me so if I reach late my timecard right."

As a result, the workers at Amalgamated began to view the relationship of time and money in a much more clearly defined way than previous generations. Many developed the habit of converting a day's wages into equivalent hourly wages in their evaluation of employment opportunities.

While there was a long commute to Amalgamated and, once at the factory, temporal discipline was enforced, the scheduling of the shift work allowed those who worked there to engage in other occupations as well. The early shift lasted from 6 A.M. until 2 P.M., and the late shift lasted from 2 P.M. until 10 P.M. Some Amalgamated workers from Anamat took advantage of this scheduling to maintain their land, since they would either work the land during the early morning or early evening anyway, in order to avoid the midday sun.

### Temporal Ideas from the United States

Migration to the United States influenced Anamatians' ideas of time during this period. In 1965, the United States changed its immigration laws to allow more migrants from the West Indies. At this time, Anamatians were among the Trinidadians who began to migrate to the United States in large numbers. Once various family members settled there, this enabled others to migrate as well, because of the family reunification emphasis of the American immigration law (Ho 1991:25).

In discussing issues of time, members of the generation who obtained their first jobs during the Oil Boom differ from previous generations in that they frequently compare Trinidad to the United States. The nature of their comparison differs from that of the old heads, who compared Trinidadians to Americans based on experiences with the American military. The generation that matured during the Oil Boom spoke authoritatively about the United States and about ideas of time found there. This authority is based on conversations they have had with relatives who live in the United States who provide advice on migration, and friends who periodically return who talk about their lives in "foreign." For instance, Kendrick, a Creole man who had worked for Amalgamated, explained,

> Like if, you go America, you jus' working. It have more work than play. Down here, you playing right through. [In] America, like, work is important. Down here, [for] the majority of people work [is] not important. You find when people finish work down here—eight o'clock. They go to work seven o'clock and they finish eight o'clock, and they getting pay for eight hours. America is not like that. You get paid for [an] hour, you understand.

Kendrick's girlfriend's sister's husband lived in the United States and periodically returned to Trinidad to visit his wife (his wife had not received a visa at the time of my fieldwork). Whenever this man returned, he and Kendrick limed together. Kendrick was anxious to learn about the United States, because he, too, wanted to move there. His visiting liming partner would say, "In the United States, you working all the time, but you make real money."

Based on such reports from emigrated friends and relatives, the image of the United States evolved into one of economic opportunity, hard work, and temporal discipline. Many Anamatians recall that, during the Oil Boom, they had the opportunity to move to the United States but chose not to, because they felt they had more freedom in Trinidad, including more temporal freedom. Their image of the United States was of a place where one worked all the time and had no time to lime.

While the Oil Boom period saw burgeoning government employment, it also benefited small-scale farmers. Harban's son reports that he did not go to work for anyone else during the Oil Boom. Instead, he sold his oranges on the streets and "made good money." "Since nobody in agriculture in those days, what you grew come like gold," he says. Indeed, the high wages in government ministries and industries drew many agricultural workers away from Anamat. This generated an acute labor shortage in rural areas, driving agricultural wages higher. Large estates had severe problems in obtaining the labor needed to work their holdings, but small-scale farmers, relying on family labor, met their labor requirements and obtained work on the estates, if they desired. The cost of land plummeted as more large estate owners abandoned and sold their land. The cost of food increased, since the number of growers decreased. Agricultural markets became sellers' markets, and real estate markets in agricultural areas became buyers' markets, allowing those who continued to work as farmers to increase their profits, their savings, and their land holdings. On the other hand, the government increasingly ignored the agricultural sector. This provided little incentive for agricultural development and made farmers increasingly antagonistic to the PNM, the party in power.

Some women took advantage of the Oil Boom and educational opportunities to learn occupational skills. These women filled jobs created by increasing consumerism during the Oil Boom period. Two of the more common jobs these young women from Anamat obtained were as bookkeepers and pharmacists. The bookkeepers worked in the stores that sold clothes or appliances that proliferated during the Oil Boom. The pharmacists found work as a result of the burgeoning private medical practices that emerged alongside the government's socialized medical system: many people making high wages preferred to see private doctors rather than go to government hospitals and clinics. Cynthia was one of those individuals who left Anamat to live near her secondary school. She was very successful in secondary school, receiving several "passes." She then entered a program to train to become a pharmacist and eventually permanently settled near the school she attended.

The Oil Boom, then, created a great rift between those who pursued agricultural occupations and those who did not. It also increased the occupational diversity found in Anamat. Temporal conflicts were no longer just between employers and employees, but between family members and friends who limed together and occasionally worked together to farm or hunt. The tension between industry and agriculture entered the home. Samuel, the foreman at Amalgamated, had to manage his work in the factory and help his father cultivate ten acres of cocoa land. His brother had left Anamat to work as a tailor. The different labor patterns of Indians and Creoles that had become institutionalized generations earlier, in which Indians emphasized family and Creoles emphasized individual accomplishment, became further embedded in the differences between the two groups. Because of these patterns, Indian secondary school students often took advantage of agricultural opportunities offered by their families. Some worked to enhance family earnings by taking nonagricultural jobs. Creoles tended to strike out on their own, leaving school early and seeking wage labor. For those who obtained work in manufacturing, time became increasingly valued in terms of hourly wages. For those involved in agriculture, the total earnings were more important than the earnings per hour. Both groups, because of schooling, emphasized temporal discipline, but their association of such discipline with earnings differed. This would have repercussions of social relations and wage labor during the recession that followed the Oil Boom, when those who had worked in factories returned to Anamat to farm.

### Declining Patronage

By the middle of the 1980s, the price of a barrel of oil started to fall from around U.S.\$26 to U.S.\$9, and government revenues fell correspondingly (Ryan 1989b:31; Ramsaran 1994). Consequently, the government began to cut many of the infrastructural improvement programs. Often, these had done little more than employ people in a form of political patronage (Thomas 1988:293). Prime Minister Eric Williams died in 1981, and George Chambers took over control of the PNM-led government (Ryan 1989a:243–90; Sutton 1984:69–72). Chambers and the PNM won the 1981 parliamentary elections, but by the next elections in 1986, the PNM had little reason to be optimistic. In 1986, there was a large government deficit, tax revenues had fallen 23 percent, and unemployment was rising (Ryan 1989b:31–32). Among Trinidadians, there was great frustration with the PNM government (Ryan 1989b; Yelvington 1987).

Like much of the rest of Trinidad, Anamat voted for a change of government. In the 1986 elections, they put a new party in power, the National Alliance of Reconstruction (NAR). The NAR consisted of a coalition of ex-PNM members and members of the United Labor Front, the party that followed the DLP in being labeled as the "Indian Party." The NAR briefly managed to bridge ethnic differences and captured a dominating majority of thirty-three NAR seats to three PNM seats in parliament.

The NAR inherited a bad economic situation and negotiated with the World Bank and International Monetary Fund (IMF) for assistance (Hintzen 1994:70–71; Ryan 1989b:258). It obtained a loan with IMF-imposed conditions, which included raising revenues through a value-added tax of 15 percent and lowering expenditures through pay cuts to the public service. Both moves were extremely unpopular. Those who worked for Amalgamated or for the Public Service during the Oil Boom found themselves unemployed. Amalgamated retrenched most of its workforce in three waves, starting in 1982. Many of the laborers from Anamat returned to squatting and "making garden" to survive. Some resorted to seasonal agricultural labor that paid much less than their former wages. With no plantations active in Anamat, local employers were mostly small farmers who needed seasonal workers to supplement their families' labor supply to pick cocoa and coffee and to keep the spaces between trees clear of dense bush.

On one morning during an orange harvest in 1990, a group of unemployed Creole men, about half of whom had worked for Amalgamated, were liming in the part of the village known as "block one hundred." A pickup truck with an Indian man at the wheel pulled up. The man leaned out his window and asked if anyone wanted a "little day's work picking oranges."

One of the Creole men queried, "What you paying?"

"Forty dollars," the man replied.

"Forty f—dollars, nah man," the Creole said, and the Indian man shrugged his shoulders and drove away. He passed by the block about thirty minutes later with three Indian men sitting in the bed of his truck.

Meanwhile, the Creoles had been discussing the job offer. "When that man say a day's work, he means all f—day, and he ain't even givin' you sometin' to eat." Another added, "Yeah, man, forty dollars is a morning's work, not no day's work, but he found some of his own to do the work, but I ain't that desperate."

While I never had an opportunity to talk with the Indian men hired that day, I did talk with an Indian laborer who willingly worked for forty dollars a day plus a meal and two cigarettes. His primary concern was the money, not the amount of time it took him every day to earn it.

Even as late as 1990, many of those who had lost work at Amalgamated still tried to figure out what to do. Some of their peers had used the severance pay to buy cars that they worked as taxis, but these cars became burdens, as they required money to keep them repaired. Other men relied on temporary labor provided by the government. In the district, either the County Council or the Ministry of Infrastructure and Public Works would begin a project that required temporary workers for either five or ten days. In several cases, the men began "making garden" by squatting on land at the edges of the teak plantation. In other cases, they relied on their families for food, shelter, and money.

Women were affected by the economic downturn in different ways. During the Oil Boom, they often continued to "make garden" while their spouses or male relatives obtained high-paying wage labor. Some women pursued their careers. In many cases, after the Oil Boom, these women retained their jobs as bookkeepers and pharmacists, making them the major wage earners in their households. In addition, women and men worked side by side in "making garden."

In most cases, the pressures of juggling several jobs as a result of occupational multiplicity changed. Formerly, it was a matter of managing a full-time job with other sources of money. During the recession, this changed to trying to be in a position to take advantage of any source of employment. Some Anamatians have become successful at this, while others have not, as will be discussed in the next chapter.

## Conclusion

Trinidadians possess ideas of time that reflect the conditions of their first employment. The oldest of the old heads grew up when plantations dominated the district. They base their laments about the "idleness" in contemporary Trinidadians on comparisons to the temporal discipline instituted by the overseers and drivers on these plantations. Slightly younger old heads use their experience on the American military bases during World War II as the basis for their discussions of time, lamenting about a "lack of dedication" in youths. Those who obtained work in the late 1940s and 1950s refer to the teak plantation, small-scale agriculture, and the anti-malaria campaign. All three of these served as forms of government patronage, with an associated emphasis on temporal flexibility. With the election of the PNM in 1956, the patterns of patronage changed, and so did the experience of the relationship between work and time. These individuals describe ethnic tensions manifested in competition for jobs, but they, too, use their experience to complain about the present. The people who obtained their first jobs during the Oil Boom reflect some of the institutionalized differences between those who emphasized agriculture and those who emphasized industry. The former are concerned with the money earned but not with an hourly calculation of wages, while the latter view is typical of those whose work lives began under the surveillance of the time clock and shift work. In addition, this era saw the beginnings of comparisons between ideas of time in the United States and ideas of time in Trinidad.

Therefore, generational differences are an important dimension to temporal differences in Anamat. A close examination of these differences also reveals the process by which ethnic groups were defined through the different ways in which their labor was commodified. These two dimensions of temporal difference continue to influence social relationships in Anamat.

# 2  Producing Times

Long before sunrise on a typical weekday morning, Anamatians who are "making market" load their pickup trucks (called "vans" in Trinidad) and begin to drive "down the road" toward the Eastern Main Road. The few individuals who work the morning shift at factories outside Anamat, and those who have other jobs that start work early in the morning, "get a drop" to Viego Grande in these vans. In Viego Grande, they know they will find taxis to take them to Arima, where they can obtain transportation to Port of Spain, the capital city. Some who "make market" travel to the wholesale market of Port of Spain. Others work at stands along the Churchill-Roosevelt Highway, a divided highway that runs from Port of Spain to Arima. Still others sell their produce at the markets in Arima or Viego Grande. Soon after the vans leave the village, taxis begin making their runs between Anamat and Viego Grande. Their first trip usually includes the same passengers day after day—people who work outside Anamat and, until 1995, secondary school students who were trying to get to their schools on time. As the sun rises, the workers for the Ministry of Public Works, the County Council, and the Forestry Department report, and farmers go into their fields in order to get as much work done as possible before the onslaught of the midday heat. Schoolchildren walk to the local primary schools, and those who commute to Anamat from elsewhere arrive—the teachers and the foremen and supervisors for the teak plantation and the Department of Works and Infrastructure. The taxis continue to make their trips back and forth to Viego Grande. By about midmorning, activity on the road subsides. Shortly before noon, people begin to return from their fields, and those who complete their tasks for the various government jobs come home. Those working the afternoon to evening shift at factories quickly finish their meals and begin looking for taxis. Others settle down to watch "pictures"—soap operas imported from the United States. As evening draws close, men congregate at the rum shops to lime. By early evening, the road is deserted again, as most people go home, watch television, and prepare for bed.

There is no major employer of all Anamatians as there was during the peak of the plantation era, during World War II, or during the peak of government patronage during the Oil Boom. Instead, both men and women participate in a variety of agricultural activities and supplement their income with temporary work for the government, whenever such work is available. Many men try to earn additional money through a wide variety of other skills, such as hunting, masonry, carpentry, auto repairs, and woodworking, to ensure a steady income. A few women cook food for sale, drive taxis, or run small shops that sell candy, but the occupational multiplicity found among most women in Anamat involves government wage labor and agriculture.

One Saturday morning, I visited Ranjit, one of several farmers I relied on for candid comments about cocoa farming. I first went into his small shop, where I found his wife working. She told me that he was in the house. After I called out, he invited me in. I found him sitting in the midst of three boxes. One contained small plastic bags, another contained candied mangoes, and the third contained small bags of candied mangoes. Ranjit, with his hands stained burgundy from the color of the food he was packaging, explained that every Monday he made his deliveries of mangoes and other foods he processed to small shops throughout the region. Soon after I sat down, he called out to his wife, "Bring Mr. Kevin a sweet drink." He and I then began to talk about what he called "the state of agriculture in Trinidad." He complained about the high cost of labor and the low prices he obtained for his cocoa. He said that the work was hardly worth the money he earned from selling his cocoa, and he threatened to cut down his trees and replace them with orange and grapefruit trees. "Let Hershey get the cocoa to flavor their chocolate from somebody else," he grumbled. "I don't make enough from cocoa to survive. I need this shop and my little business," he added as he laughed and pointed to the boxes surrounding him.

Ranjit's life involves the intersection of multiple cycles: the commercial cycle of delivery of the candy he and his wife make, the agricultural cycle of harvesting and processing cocoa, and the daily cycles of business at his shop. Ranjit's household is like most others in Anamat, because of the intersecting cycles of agriculture, other forms of labor, and commerce. Many households also have cycles pertaining to child care. In this chapter, I discuss the different cycles in which Anamatians participate.

## Domestic Tasks

The burden of cooking, cleaning, and child care rests upon the shoulders of women, many of whom work, as well, in unremunerated agricultural work or wage labor. Indeed, it is a common pattern in Trinidad for women to manage both domestic responsibilities and wage work (Harry 1993; Reddock 1993, 1994; Yelvington 1993b, 1995). This does not lead to a distinctive "women's time" but instead points to the complexity with which times and rhythms are gendered. As Barbara Adam argued, the "differential treatment of times becomes visible in the sequencing and prioritizing of certain times and in the compromises in time allocation that have to be achieved on a daily basis" (1995:95). In the relationships between men and women, these different priorities, and I would add different rhythms of activities, lead to negotiation, cooperation, or conflict. Importantly, however, this involves a playing out of ideas concerning gender as well as time.

Basmi, an Indian woman, exemplifies the pressures women face when managing these responsibilities and the extent to which coordination of household activities embodies gender and age relations. Her household includes several of her children, some of her children's children, and her son's wife. She makes sure the small children are cared for; she coordinates the efforts of the older children in "making garden"; she has meals ready for her husband who has a regular work schedule on weekdays, and for other family members who do not live with her but who sometimes eat at her house; and finally, she runs a small shop catering to schoolchildren. To coordinate the lives of the fifteen or so people who rely on her, and who either live with her or nearby, is a daunting task, but one that she has done for many years. Often this coordination constitutes the day-to-day relationships Basmi has with the people in her household. If one of her children does not complete a chore when Basmi needs it completed, it is a source of conflict. In recent years, she has expressed gratitude for the help received from her daughter-in-law and her adult daughters in coordinating the household activities. Her adult sons and particularly her husband provide the greatest threat to Basmi's efforts at temporal coordination. Her husband's job is relatively inflexible in its scheduling, whereas her sons either "make garden" or work as day laborers for other farmers. Her need to coordinate around her husband's job and her ability to negotiate with her sons reflect not only the different types of work in which they engage but also issues of gender. Since her attempts to work around her husband's activities and negotiate with her sons extend to weekends and holidays, this indicates that more is involved in Basmi's efforts than the rhythms and demands of particular jobs. It also indicates a different character of gender relations between a mother and her sons and between a husband and wife in Basmi's household. There is also a weekly instance when the members of her household, including her husband, must adapt to her activities. Every Saturday she meets one of her sisters in Viego Grande to "market" together. This event is as much a social occasion for them as a household necessity.

Cooking, washing, cleaning, and child care are gendered activities that involve a great deal of effort, and they must be coordinated with the activities of other household members. When those other members are adult males, it becomes clear to what extent conceptions of gender are related to issues of time. The preparation of meals takes time, particularly since cooking many Trinidadian dishes is labor-intensive, and most Anamatian men express an aversion to "fridge food"—in other words, leftovers. Even so, women share with one another the techniques they use to prepare food ahead of time, keep it in the refrigerator or covered, and then serve it as if freshly cooked. While a few households have washing machines, most laundry is done by hand. When there is an electricity blackout or when the water supply is insufficient to run the washing machines, there is no choice but for every household to wash clothes by hand. Keeping a house clean is a constant challenge and is necessary to prevent the onslaught of insects. Small children must be supervised constantly, which generates additional demands on anyone who must also cook, clean, and wash. In many cases, the completion of these tasks involves the elegant orchestration of several people, many of whom also have other jobs.

A topic of discussion in the village is the success, or lack thereof, of households to manage these responsibilities. For a household to be deemed "respectable," the children must be kept clean and well groomed, and they must arrive at school on time. These evaluations are important not only in the cultural constitution of gender roles but also in terms of who makes these evaluations. The public appearance and behavior of children and how they reflect on their mothers are most commonly discussed by adult women and by young adult men liming by the roadside. Indeed, from the men's roadside discussions, it seems that the care for children is a major indicator to them of how well the household is run. Every morning when school is in session, the children walk through the village to get to the elementary school. Every morning, a group of young adult men congregates at the end of the trace leading to the elementary school. Their purpose for being there is that it is an ideal place to obtain information about potential employment or that it is on the way to their gardens. The passing of schoolchildren provides these men with opportunities to comment on the children's parents, particularly their mothers. Children who are still "on the road" after school starts or whose school uniforms are dirty or wrinkled inspire negative comments from the men.

Women's roles at the center of temporal coordination in households suggest an important source of power over the lives of others. This does indeed seem to be the case with children and younger siblings. This conceals the deference wives pay toward their husbands' nondomestic activities and the extent to which mothers are evaluated in terms of their abilities to run their households.

### Agriculture

Agriculture is the most important money-making activity in Anamat. It takes several forms, ranging from the cultivation of food crops in "gardens" to the production of cash crops on large estates. Many gardens involve squatting on government land. Squatting allows those who do not own land to grow their crops without paying for the use of the land. Squatting's limitation is the inability of a farmer to cultivate the same plot of land for long periods of time. This prevents squatters from raising lucrative tree crops. A few enterprising squatters eventually purchase land with money they save. Others, through marketing skills and saving, acquire enough money to buy a small, used van (pickup truck). They use the van to take produce to market. They generate extra income by providing transport services to others for a fee. Some choose to spend money on other things, such as the education of their children, rather than attempt to save money to remain in agriculture. Many, however, find themselves so enmeshed in small debts and living costs that they remain trapped in their economic situation.

As in the past, for many who rely on agriculture in Anamat, it appeals because it provides independence. For these people, such independence balances the low prestige that agricultural work currently holds among Trinidadians. For some, this desire for independence is based on a lack of trust for those outside Anamat. As one man said, "Whatever happens in this world, [Anamat], as the place is called, has a plentiful supply of food. I always expect a war. That's why I plant so much dasheen and cassava. When the bombs start dropping, I still be eating my ground provisions." His statement was somewhat prophetic. A few months after he told me this, there was an attempted coup d'état (see Ryan 1991a; Searle 1991; *Trinidad Express* 1990) followed by island-wide food shortages. The people of Anamat had an abundant supply of food, and many sold their produce, particularly breadfruit, at huge profits in markets throughout the island.

Anamatians view the independence that agriculture brings as a guard against political instability in the world outside Anamat and as a protection against economic instability. There is an attitude among some Anamatians that they should take advantage of whatever economic opportunities are available, but that, when times are tough, as in the recent recession, they still retain their gardens on which they can depend. During the recession of the late 1980s and early 1990s, Anamatians frequently commented to me, "Nobody starves here." They would then compare Anamat to "town" (Port of Spain). As one man said to me, "In town, there's no place to grow anything. You must hustle for your food. Here, even if you have no land, you can get fig to eat and orange to suck."

Indeed, minor theft of produce is an accepted reality among farmers in Anamat. They draw a distinction between such "thiefing" in order to get something to eat, and "thiefing" for sale. One day, a neighbor of mine went to his garden to discover that all of his dasheen, a root crop, had been dug up and carted away. As he vented his anger, he said, "I don't mind someone taking some if they hungry and have nothing to eat, but what makes me vex is digging it all up and bringing in a van to take it all away. I know that tomorrow, that dasheen will be for sale in the market."

While some people, particularly young adults, see agriculture as an occupation of last resort, many people continue to cultivate land even if they obtain wage work elsewhere. During the Oil Boom, many of the men who worked for Amalgamated also grew cocoa and ground provisions. Currently, there are many households in which one member works outside Anamat and the other members cultivate land, although this seems to be more common among Indians than among Creoles. For instance, Naresh has one adult child who is a civil servant and another who maintains the family land. Naresh now rarely works his land because he is able to divide the responsibilities among his children. For many other Indian men of his generation, his family provides a model—they, too, hope to have some children to work the land and some to obtain office jobs.

While agricultural pursuits seem to provide some degree of flexibility, they do have temporal pressures which farmers recognize. The practice of "making garden" involves the cultivation of food crops. It is the contemporary form of conuco agriculture (see Berleant-Schiller and Pulsipher 1986; Hills 1988). To "make garden," one clears land and starts planting during the dry season, which extends from March through May. The most important crops are bananas, pigeon peas, and "ground provisions" (root crops). Ground provisions take about nine months to mature, so for those who plant during the dry season, they begin to harvest around December. Pigeon peas are harvested at around the same time. Bananas ripen year-round.

The major tasks involved with this form of cultivation include mounding the ground provisions before the rainy season in order to prevent erosion, weeding the land, and harvesting the crops. Most gardeners begin work in their plots at sunup.[1] There are exceptions to this, particularly among those who have jobs with the Forestry Department, the Ministry of Works and Infrastructure, or the County Council. These individuals often work on their land in the late afternoon. The tasks the workers set are what determine their work rhythms. These tasks vary considerably, depending on the season and the inclination of the farmer. The tasks of clearing, planting, and harvesting require more time and effort than the tasks of maintaining the field during the rest of the year. During clearing, plant-ing, and harvesting, other activities are often of secondary importance and are scheduled around the work demands of the garden. During the rest of the year, people structure the rhythm of work to fit around other activities, such as watching soap operas, liming, or eating meals.

The advantage of gardening is that the times that it demands the most labor are frequently not the times that cocoa and coffee farming demand the most labor. Consequently, those who garden maintain their gardens and work in other forms of agriculture at the same time, without sac-rificing productivity.

Those who "make garden" sell their produce as far away as Port of Spain in both retail and wholesale markets. Their choice depends upon personal whim, the availability of transportation, and the current prices. These marketing activities have very different rhythms. Selling in the wholesale market requires leaving Anamat very early, but since the produce is bought in bulk, the amount of time needed to sell the produce is often much less than when the farmer sells the produce in retail markets. If farmers choose retail markets, they typically sell in markets closer to home. It takes longer to sell the produce, and less produce is typically sold, but it commands higher prices. For many, the choice between selling in retail and wholesale markets is based on the quantity to be sold. For others, the preference for spending time in one location versus another—say, liming in Port of Spain versus Arima—determines whether they sell in retail or wholesale mar-kets.

Those who "make garden," like peasants in other parts of the world, supplement their income by working for others (Hill 1986:109–10; Wolf 1966:45), although the nature of the work varies because most squatters are not solely farmers or even unskilled laborers. During the years of the Oil Boom, from the mid-1970s until the mid-1980s, when the high world price of oil buoyed the Trinidadian economy, most agriculturists learned skilled trades such as carpentry, masonry, welding, auto-body painting, and mechanics—in many cases funded by government patronage (see R. Henry 1988:482–87).

Among those who returned to agriculture during the recession were a small number of people who became sharecroppers, which they describe as working "half." For instance, Jerome worked "half" on his uncle's land. He gave his uncle half the cocoa and coffee the land produced, and all the rest of the land's produce was Jerome's. In addition to cocoa and coffee, the land contains breadfruit, orange, and coconut trees. Jerome remarked, "I don't get much because the land come abandon—it is like working a fresh piece of land." By this, Anamatians mean that it had not been properly maintained by having the underbrush periodically cleared and the drains kept open. Most of the sharecropping arrangements in Anamat involve such land. While sharecropping allows someone who does not own land to share in profits gained from cash crops such as cocoa and coffee, rehabilitating neglected land takes a lot of work, often for very small returns in the beginning. In many cases, it takes as long as five years to rehabilitate abandoned cocoa land. Often, landowners see offering their land to sharecroppers as a way of returning long-abandoned fields to productivity. At the same time, most sharecropping arrangements are between kin and appear as the result of an obligation felt by landowners to help their unemployed kin, and a reciprocal obligation of the kin to accept the offer, sometimes in hope of inheriting the land upon the owner's death.

Sharecropping has its temporal hazards. The worker has to satisfy the owner of the land by harvesting and processing the cash crops promptly and effectively, but at the same time, the worker is inclined to concentrate more on his own crops, usually food crops, rather than on the cash crops which are halved with the landowner. These conflicting motives can lead to mistrust between owner and sharecropper, despite kin ties.

Many Anamatians, sharecroppers included, view squatting and practicing shifting cultivation as more immediately profitable than sharecropping—particularly in recent years. Indeed, even while Jerome worked on his uncle's land, he was "making garden" on government land.

Not only is gardening becoming profitable, but in yet another form of patronage, the government has offered incentives to garden. A large teak tree plantation is located in Anamat. In the 1980s, the government paid would-be gardeners money to clear land on the edge of the plantation. These gardeners then planted their crops, and the Forestry Department planted teak trees in the same plot of land. The government then paid the gardeners to care for the young teak trees in the gardens, and it also cleared drains in the teak. These subsidies ended in the 1980s, but the government continues to encourage gardening on the edge of this plantation by clearing forest, offering the land to gardeners, and then paying the gardeners $20TT (about U.S.$4.70 in 1991) per acre once they are done working the land, which is at about the same time as when the teak saplings no longer need shade.

Those who have long-term control over land have a greater potential for earnings, greater independence than other agricultural workers, and greater control over their labor and time. Aside from land leased from the government, small farmers own most of the land in Anamat, in plots ranging in size from five to ten acres. Among the small estate owners, a few own as many as thirty or forty acres, but this is unusual, as well as unmanageable, without hired laborers. A handful of individuals own vast expanses of land, but in these cases, much of the land has not been properly managed for many years. For instance, in the 1970s, a family in a neighboring village bought the land holdings of the Stollmeyer company in Anamat. At one time, the Stollmeyers were among the wealthiest families in Trinidad and had land holdings all over the island (de Verteuil 1994:92–138). Their plantation in Anamat consisted of several large plots of land distributed throughout the district and a large barracks and cocoa house for processing the cocoa from the land. When the company sold the plantation, most of the land fell into disuse, since the family that purchased the land did not hire nearly the labor force the Stollmeyers maintained. Now, after more than twenty years of neglect, several hundred acres of former Stollmeyer land are reverting to forest.

To protect themselves against the whims of the marketplace and crop failure, most landowners cultivate a variety of crops and do so in such a fashion that during every part of the year there is something to harvest and sell. The most important crop is cocoa, which bears fruit a couple of times per year. During periods other than harvest times, cutlassing, weeding, and arrondeering are the main tasks to be performed in the cocoa groves. Cutlassing involves cutting and clearing away the undergrowth between cocoa trees. Because of the rich soil and heavy rains, this needs to be done between two and four times every year. Before cutlassing, it is best to weed in a ring around every tree—this is known as arrondeering. This prevents injury to trees during cutlassing. Besides cutlassing, it is necessary to prune the trees lightly every year, and to do so more severely every three to four years, thereby maximizing the productivity of every tree. Periodically, drains also need to be dug and maintained—poor drainage in cocoa walks promotes the rotting of cocoa pods.

Harvesting cocoa involves cutting the pods from the tree, cracking open the pods, and removing the beans. A single person can harvest several acres in a few days. The raw beans must be processed, however, and this takes time. They must be "sweated," then dried. To sweat cocoa beans involves piling them and covering them. After several days of sweating, the beans are placed in the sun to dry. Drying requires turning the beans occasionally so they dry evenly, and remaining constantly vigilant against rain. Finally, to clean the beans before sale they are "danced." Water is gently poured over the beans while someone moves through them turning his or her feet. This separates the chaff and insects from the beans. None of these activities requires constant work, but the fermenting and drying processes do require frequent monitoring. In many cases, people take care of other duties around the house while they are keeping track of their cocoa.

Farmers' dissatisfaction with cocoa as a cash crop is increasing. First, the trees are growing old and need to be replaced. To replace the trees involves a willingness to forgo significant production for three to five years, and it takes a new tree twelve years to reach peak production again. Second, the cocoa dealers pay higher prices for large grains of cocoa than they do for small grains, but many local trees are of a variety that produces small-grained cocoa. When a cocoa farmer sells his cocoa to a dealer, the dealer pours it through a sifter, and if the beans pass through, the farmer gets a lower price. But some farmers complain that the trees that produce the large bean do not produce as many pounds of cocoa per acre as the small-bean trees. Finally, under pressure from the corporations that buy cocoa, there is talk of insisting that the cocoa beans be sold before any processing, thereby allowing the corporations to process the beans. This would significantly reduce the price that farmers get paid for their cocoa. Consequently, even though Trinidad grows some of the highest quality of cocoa in the world, many of the farmers who grow this cocoa consider replacing their cocoa trees with other crops, such as bananas, papayas, and citrus fruits.

This process has already begun to occur with regard to coffee, never a very profitable crop for Anamatian farmers. Coffee rarely fetches a good price, and harvesting it takes more labor and time than harvesting cocoa. To harvest coffee properly, one should pick only the ripe, red beans and leave the green beans. Anamatian farmers have little patience for this, however, and often simply spread a tarp underneath the tree and strip the branches of all the beans, both green and red. In part because of this practice, local coffee is considered to be of a low quality, and this reduces the price for all local coffee, even the price of coffee from those farmers who carefully pick only the ripe beans. Consequently, there is no benefit from picking beans with care, and the crop itself produces relatively little income.

Thus, while cocoa and coffee were the two dominant products for the past century, both of these crops are being replaced. The farmers see three benefits from doing this. First, crop diversification protects against price fluctuations, and, second, it minimizes the threat of disease. For instance, witchbroom disease has devastated the cocoa crop several times. Currently, there is another disease infesting cocoa trees that causes the pods to rot rather than ripen and thus reduces productivity.[2] In addition, a virus is killing papaya trees. Third, diversification allows the farmers to market much of their produce themselves rather than sell it to middlemen. The farmers find the local markets more remunerative than the global markets.

All agricultural pursuits in Anamat have several qualities relevant to Anamatians' ability to coordinate their social relationships and activities. First, agriculture involves varying temporal demands throughout the year in planting, harvesting, and caring for the crops. Cocoa involves additional demands of processing. Second, the major crops—cocoa, coffee, tubers, and bananas—all allow for a great deal of temporal flexibility in their maintenance and harvesting. This flexibility allows farmers to pursue other activities. The main temporal challenge is meeting commercial schedules of marketing and transporting produce.

### Transporting Produce

Agriculture supports many other occupations in Anamat and, as a result, these occupations must be coordinated with agriculture. In Anamat, agrarian rhythms and other occupational rhythms come into conflict. Those who transport the produce experience the tension between agricultural rhythms and commercial rhythms the most acutely. For instance, if they get a cocoa farmer's produce to the buyer too late, the buyer is angry, but if they arrive too early, the farmer gets angry because he or she has to wait for the cocoa buyer to arrive. Often, farmers ask taxi drivers to check to see if a buyer is open before they venture forth. Formerly, very few Anamatians owned trucks, and those who did charged to carry several farmers and their produce to market. When there were few truck owners, the farmers were at the mercy of the truckers' schedules. With the proliferation of vans, this pressure is no longer as great, except for large loads. Growers who produce large loads must rely on the few individuals who own large trucks.

The few individuals who have invested in large trucks use them for a variety of jobs. Their most regular type of work is hauling produce to market for farmers. In this case, several farmers band together to hire the truck to take their crops to the market in Port of Spain. This most commonly occurs when farmers perceive good prices there. The best example of this in recent memory was during the food shortages that followed the 1990 attempted coup d'état in late July and early August of 1990. These shortages came at the same time as an exceptionally good breadfruit harvest. Normally, breadfruit is not a very profitable crop, but because of the food shortages, farmers were able to make large profits in the week that followed the attempted coup d'état.

These truck drivers also haul water containers for the government. While Anamat is located in a rain forest, many residents rely on truck-borne water. The village's reservoir is small, and the water supply is insufficient to maintain water pressure during the dry season. Until 1997, the Water and Sewage Authority (WASA) maintained the local water system poorly. Finally, many homes on traces have no pipe-borne water. Because of these factors, most households in the community rely, at least partially, on rainwater they collect. Every home has some means of storing water, either large tanks constructed for the purpose or fifty-gallon barrels. Those with water tanks connect the tanks to the water line and gutter systems to collect run-off from their roofs. Even this is often insufficient to provide a water supply during bad dry seasons. As a result, quite a few homes supplement their rainwater supply with deliveries of water by trucks provided by the government—yet another manifestation of patronage. The truck drivers who work for the government have large, cubical water containers loaded into the beds of the trucks and then have these containers filled with water. Officially, the County Council water trucks deliver water only inside the traces and not along the main road, yet those who live on the road require water, as well. They rely on unofficial deliveries of water from County Council truck drivers who are kin or friends, or on others who own vehicles and who are willing to carry water containers and fill them where there are stand-pipes for the pipe-borne water. These trucks come on designated days, and people arrange their routines to ensure getting water.

For truck drivers, the peak periods of business occur during the dry season and at the time of harvest. Truck drivers find it difficult to coordinate their trucking with cultivating their land. For this reason, no truck drivers engage in the production of root crops, although a few do work cocoa land shared with other family members. One such driver is Bertram. He has been driving trucks for years. Recently, when his father died, he inherited a large amount of land. Bertram relies on hired labor to maintain the land, but because the land is well maintained, it is profitable.

Many Indians who own pickups use their trucks to bring imported bananas from the wholesale market in Port of Spain to Anamat, to ripen for sale in retail markets. To do this requires the ability to transport large quantities of green bananas from the market in Port of Spain and then to transport the ripe bananas to the locations where they are sold—either along the highway, on the streets in Port of Spain, or in various retail markets. Ripening the bananas requires refrigerated rooms. This enterprise faces several challenges. One obstacle in buying additional bananas is the capital necessary for the vehicle and the facilities to ripen the bananas. Just as important as the vehicle and the facilities is the need of several trustworthy workers to maintain family land, buy and ripen bananas, and market them. The only families that engage in this are those with enough male kin to cover all these responsibilities.

Trucking and transport emphasize temporal flexibility, as does agriculture. Such mutual flexibility makes it possible for the two vocations to function together, but truckers have to be sensitive to commercial rhythms outside Anamat. After all, Anamatian trucking depends on agriculture and markets. Indeed, different members of a household or kin group usually perform domestic tasks, trucking, and agriculture. In this way, their efforts complement one another, and there is minimal temporal conflict between their tasks.

## Government Work

Most government work in Anamat also exhibits temporal flexibility by providing opportunities for workers to have enough time to tend to their land holdings. The government pays permanent workers according to either task work or day work. A task is roughly equivalent to a day's work, but if the worker finishes ahead of time, he or she can go home. Day work generally begins at seven in the morning and ends at three in the afternoon. In both task work and day work, workers have plenty of daylight left afterward to work their land or complete domestic tasks. This differs from the task and wage work of the plantation period that involved longer work hours with little temporal flexibility. Government and plantation work resemble one another, however, because both involve patronage and not just concerns with productivity. The plantations served as patrons for small farmers whose labor supplemented the plantations' resident workforce, and the government uses wage work opportunities to garner political support.

For most Anamatians, government work takes the form of temporary work amounting to five or ten days at a time. The major government employers in Anamat are the Forestry Department, the Ministry of Works and Infrastructure,[3] and the County Council. During the recession of the 1980s and 1990s, only one project requiring temporary workers was run in Anamat per year, so these temporary workers had to seek other forms of employment for the remainder of the year. In addition to public works, the government teak plantation is a major employer in Anamat. As mentioned earlier, the same land on which squatters plant their crops becomes, after the squatters move on, an addition to a large teak tree plantation. The squatters' plants shield the seedlings and protect against erosion, and by the time soil fertility declines enough so that the squatters shift to a new site, the trees are mature enough to survive. Many of these farmers work for the Forestry Department in the teak plantation.

The Forestry Department in Anamat relies mostly on permanent workers, rather than on "casual" (temporary) workers. Even so, it involves patronage relationships. The overseer position of the teak plantation has been in the same family since the origin of the enterprise. He has a great deal of responsibility in hiring and organizing the workers. Most of the workers are Creole, but this is not the result of overt racial discrimination.

Indeed, some Indians have worked in the teak plantation, and a few continue to work there, but many Indians reported that they did not like the work, and others have chosen not to apply to work there, assuming that they would not like the work.

In Anamat, Forestry Department jobs involve maintaining the large acreage of trees in the teak plantation. This plantation grows considerably every year, and a staff of workers maintains the trees and selects the trees to cut. Teak trees reproduce by sending out shoots. Because of this characteristic, forestry workers must choose which trees should be cut from among the different generations of trees found in any one spot. This maximizes production by keeping the proper spacing between trees and the proper mix of old and young trees. After fifteen years, the trees are suitable for posts. After twenty-five to forty years, they are suitable for furniture and are sold—often to Scandinavian furniture producers. TANTEAK, a government-supported business that cuts and transports the trees, brings large trucks into the village when the trees are harvested. The heavily loaded trucks destroy the road, which lacks a base substantial enough to withstand the traffic. This, combined with the geology of the soil underneath the road, promotes land slippage and generates work for the Ministry of Works and Infrastructure.

Of all the major employers in Anamat, the teak plantation provides the shortest regular work hours. The work is hard, and workers suggest that this is one rationale for assigning work according to tasks that many complete by midmorning. In many cases, the workers finish their work much closer to their gardens, which are located on the fringes of the teak plantation, than their homes.

An especially important source of government patronage in the form of jobs is the Ministry of Works and Infrastructure. This ministry employs both casual and permanent workers. Most adult villagers, both men and women, have worked for the Ministry of Works, at least as casual workers. Permanent workers labor all year on maintenance, as well as on large projects. Permanent workers fill potholes, dig drains, and keep the foliage on the berm cut down to improve drivers' visibility. Large projects require additional workers, a need that the Ministry of Works meets by hiring casual workers. These projects usually take place around Christmas, before Carnival, and during the period immediately before major elections. The timing is intended to "give people a little cash" around the holidays, and the jobs are clear examples of political patronage taking advantage of the season. The timing of major temporary hires before Christmas can place a burden on those who cultivate ground provisions, since their period of harvest corresponds to the period of the temporary government jobs. One advantage such farmers have is that root crops can be left in the ground after they mature to be harvested when marketed. This allows those who obtain temporary work and are cultivating ground provisions to coordinate their gardening with the government work to earn some extra cash— they work one or two weeks for the government, and then they harvest their crops.

The use of this ministry for political purposes is notorious in Anamat. Major projects usually begin in the months before parliamentary elections. These elections also typically occur during the Christmas season. Anamatians recognize such major projects as feeble attempts to influence votes. In the 1995 elections, the matter was made even worse when a project started very shortly before the elections, and a rumor spread that the project would hire only supporters of the People's National Movement (PNM), the political party in power. Many villagers, including prominent PNM supporters, openly rejected this position and pushed for the project to hire anyone, taking the long-established Anamatian position that Anamatians are more important than any outside agency or group. While it is unclear that the PNM lost any votes because of the rumor, most of those who reported that they voted for the PNM said they did so out of friendship with the PNM candidate, and despite the PNM's "stupidness" in the Public Works project.

Moreover, the system of hiring encourages favoritism. Potential workers present their identification card and have their names placed on a list, from which are drawn the names of those who will work for the project. Favoritism often plays a role in the selection of names from the list. Political party affiliation, ethnicity, and kinship influence one's chances of getting work. In this way, hiring temporary workers becomes a means of both political patronage and a source of political dissension.

The Ministry of Works and Infrastructure only cares for the major roads; it does not maintain the many traces that lead away from the main road into private land. Traces are the major means of access into the agricultural land far from the road, and they also serve as shortcuts for those on foot, by crossing the rugged terrain in a straight line; the road, on the other hand, meanders around hills and tends to follow the river. Many years ago, most of the population of the village lived along these traces, but with increased amenities along the road—electricity, running water, and improved taxi service—people building homes have tended to rent or lease land along the road on which to construct a home, and then to use the traces to walk to their land.

In effect, the County Council does the same work as the Ministry of Works, and is, in fact, supervised by the Ministry of Works. The bureaucratic difference between the two is that the Ministry of Works and Infrastructure is run by the national government, whereas the County Council is run by locally elected officials. With regard to the organization of work and hiring practices, there is not much difference, except that the County Council rarely tackles jobs with much complexity in engineering or jobs requiring heavy machinery.

The County Council is also a source of patronage for the village. In those years when the County Council representative is from Anamat, the County Council initiates major projects, such as the construction of the village's cricket ground with covered, concrete stands. When the representative is from another community, there is noticeably less County Council work in Anamat.

### Occupations Complementary to Agriculture

Between the money earned from agriculture and the wages obtained through working for the government, there is a small, but steady demand for service industries in Anamat. Most individuals who provide important services in Anamat also own land that members of their family cultivate. The village has several small grocery stores. These small shops serve as general stores, supplying villagers with food, supplies, and other items such as hoes and cutlasses (machetes). Before recent improvements in transportation, such stores served as the major link for receiving goods from the outside world. Even now, for many who do not have a car or the money to take a taxi to the nearest market town on a regular basis, these small stores are the primary supplier of everything they need. Another important feature of these shops is their flexible hours. On the one hand, they often have irregular hours. On the other hand, shopkeepers complain that customers come by at any time of the day or night. Yet it is the temporal flexibility of the shops, as well as their accessibility, that maintains their clientele.

The residents of Anamat rely on several other occupations to supplement their income. Some men earn extra money by modifying cutlasses. Anamatians prefer to use cutlasses with the point cut off and both edges sharpened. In addition, to cut grass, whether lawns or the bush among their fruit, cocoa, or coffee trees, they bend the handle end of the cutlass blade and mount it to a long handle. Several men in the village specialize in making handles and modifying cutlasses according to the owners' specifications. As is often the case with custom work, the craftsman and the customer conflict over when the job is supposed to be complete.

Although the construction business is not prosperous in Anamat, several men also work as part-time carpenters and masons. They have built most of the homes in the village. Construction professions, too, have temporal irregularities. Most homes are built piecemeal, often remaining unfinished for years until the owner can afford the materials and labor to finish the job. As a result, village carpenters and masons do not count on this sort of work for a steady, predictable income. Most of them also farm.

The coordination of construction becomes a daunting task, even once one does have the money, as I found out. I wanted to replace some of the termite-infested floorboards in the house I rented. This ended up being a multi-week process. First, I had to arrange with a carpenter to do the work. The man I asked was also busy in his gardens and marketing his produce. He and I agreed that it would be best if he came with me to buy the wood. To do this, we needed to arrange with a man who owned a van, so we would have the means of hauling the wood back to Anamat once it was bought. This man drove a taxi and arranged for his oldest son to drive the van for us. This son cultivated a garden and was finishing up work on harvesting and marketing ground provisions. Consequently, he was harvesting and marketing at the same time that he worked for his father, whom I had hired to help the carpenter to get the lumber. After several days, everything fell into place, and we agreed to a time to meet and travel to lumberyards and sawmills for the lumber. Naively, I scheduled a time for us to meet.

The time came and went. A friend who was with me said, "Don't study that. He comin' jus' now." Sure enough, the carpenter I hired appeared about forty-five minutes later. He and my friend then started joking about how I was probably thinking he was not coming at all. Meanwhile, the son of the owner of the van passed and called out, "I goin' up the road to come back," implying that he was going home first. What I thought should be a twenty-minute round trip, having driven it myself to make the arrangements, turned out to take another hour—he had wanted to bathe after coming out of the garden, before he traveled to town with me. By the time we brought the lumber back, the carpenter said it was too late for him to start work and that he would come back another day. After witnessing other such projects, I realize in retrospect that what was atypical about my attempt to get my house fixed was my fixation on the clock and my ignorance of the work demands and routines of others.

Furniture making is another important business in the village. Many people grow hardwood trees amid their cocoa and coffee trees. A "joiner," as furniture makers are called, purchases a tree for some wood and then makes chairs and sofas to sell to furniture stores as far away as Port of Spain. The business is not as lucrative as it used to be, and only one man now makes joinery his primary income source. Joiners complain about the small amount of money one makes after paying for transportation and materials, and the relationship between joiner and store is full of temporal tension. Many stores follow a monthly schedule of stocking, billing, and payments, but the joiner and those who transport the furniture want immediate payment. For those who transport, their expenses and time demands follow agricultural cycles. Still, a handful of craftsmen do make furniture for special orders.

Taxis provide one of the most important services in Anamat. They make the outside world readily accessible to Anamatians. In Trinidad, taxi drivers do not just carry people: they run errands for the sick, carry produce, and deliver messages (Abdool 1979). Anamatians value this latter function, since there are no telephones in the village. The taxis run from Anamat to Viego Grande. From there one can find transportation to many other parts of Trinidad, including Port of Spain, the capital and largest city. Anamat is the terminal point for the taxis who ply the road between the village and Viego Grande. Every day, the taxis begin work early in the morning, to serve the commuters and schoolchildren during the week, and those going to market on the weekend. Some work long after dark to ensure that all the villagers get home.

All of these subsidiary occupations emphasize temporal flexibility, in part because of the tension of scheduling agricultural interests versus commercial interests. In the case of shopkeeping and crafts, the occupations allow for those involved to work around the house at the same time they "mind the store." When a customer comes, the customer calls out, and the proprietor responds, "Coming jus' now," stops whatever he or she is doing, and then cares for the customer. Other trades and vocations follow the pattern of truck drivers, where the necessary tasks in the home are distributed according to the differing rhythms of the household's members. In all cases, work within Anamat emphasizes temporal flexibility.

### Working Outside Anamat

Jobs outside Anamat often have less temporal flexibility than those in Anamat. Several Anamatians work in civil service and teaching jobs. The few civil servants commute from the village to Port of Spain every day, choosing to live near kin rather than their place of business. Many of the teachers in the two primary schools in the village come from the village or have kin in the village. Some families are heavily involved in local education. For instance, at one point, one family contributed approximately one-third of the teachers in the village.

Several residents work in factories near Arima, about twenty miles away. These workers are fewer in number than they were during the Oil Boom. As during the boom, shift work is common—the first shift lasting from six in the morning until two in the afternoon, and the second shift lasting from two in the afternoon until ten at night. Several Anamatians also work at Piarco International Airport, located on the other side of Arima from Anamat. Both factory workers and airport workers have places to stay outside Anamat when they cannot find transportation back to the village late at night. In order to get to work for the morning shift, these workers obtain rides from drivers going to market early in the morning. In this way, the shift workers get to Viego Grande early enough to find taxis to Arima and then to work.

In general, the jobs outside Anamat require greater time discipline than the jobs in Anamat, and in general, the outside work is more closely governed by the clock than the jobs in Anamat. Teachers and factory workers, in particular, must contend with very specific definitions of time. In the case of teachers, they must also define time for their students and inculcate in their students temporal awareness and discipline (Birth 1996). In the case of factory workers, they must contend with time clocks and with managers who are concerned about the relationship of productivity and time.

## Occupational Multiplicity and Time

Since many individuals work in several occupations, and since each occupation has its own rhythm, the problems Anamatians face in coordinating their activities and relationships are substantial. Despite the expectation of temporal chaos in such occupational multiplicity, Anamatians act as if there is order. True, the diversity of rhythms generates unpredictability, but Anamatians manage through temporal flexibility. Such flexibility is characteristic of many activities in Anamat.

Work patterns powerfully structure daily rhythms and the workers' temporal coordination of social relationships. Despite the close association of agricultural production and natural rhythms, however, the commonly held assumption that peasants' close relationship to nature homogenizes their conception of time is mistaken. Nature manifests too complex a system of interwoven rhythms to produce a single human model of natural times. Farmers do not respond to nature in its entirety, but only to specific dimensions of the ecological niche they occupy. Different types of farming, and consequently different ecological niches, result in different conceptions of time emerging from agricultural practice. Cocoa cultivation depends on the seasonal rhythms of cocoa trees that produce two major crops per year. The activities of caring for the trees are determined by two factors: first, the rhythm of harvests, and second, diurnal rhythms that determine when tasks can be performed with maximum efficiency. The clearing of grasses and other undergrowth is best done in the morning or evening, when the plants to be cleared are "softer," meaning easier to cut. In contrast, those who "make garden" closely tie the cultivation of tubers to the ability to clear land by slash-and-burn methods during the dry season. Tuber crops require the most work during planting and harvesting. Cultural interpretations of the effects of lunar phases on plants determine planting. Local wisdom holds that different plants have different reactions to particular phases of the moon. The determination of when to start clearing land for burning is based on yet another plant, immortelle trees. One variety of immortelle blooms at the end of the rainy season, thereby signaling the approaching dry season. The other variety of immortelle blooms during the dry season, and it signals the end of the dry season when it starts to lose its blossoms. Consequently, the annual agricultural cycle for cocoa growers arises in response to the maturation of cocoa pods, whereas the annual agricultural cycle of those who make garden emerges from interpretations of the relationship between immortelle trees and the dry season.

Other occupations have associated times arising in relationship to extra-village social factors. School schedules fall into this category and have repercussions beyond the lives of students: families that rely on their children's labor must accommodate school; taxis that transport children organize their day around increased demand just before and just after school hours; and stores that sell candy open for those times when children are going back and forth to school.

In summary, the influence of natural factors and extra-village social factors generates multiple temporal rhythms. The link between occupations and time generates temporal conflicts because of Anamat's occupational diversity. Anamatians interpret these conflicts in terms of the dimensions of difference found in Trinidad: sometimes ethnicity, sometimes gender, and sometimes age or kinship becomes the basis for overlooking temporal conflicts, or for explaining them.

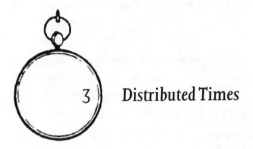

3  Distributed Times

Cultural ideas, including those concerning time, are distributed among individuals, statuses, and roles (Hannerz 1992; Shore 1996; Swartz 1991). All individuals have personal histories in which they have learned specific cultural knowledge, such as the different models of time found in Anamat's different generations. Roles, as constellations of expected behaviors, are extremely important in the social organization of individuals—the topic of this chapter.

People possess immense amounts of cultural knowledge, but only a small portion of this knowledge is consciously applied at any given moment. For instance, many parts of the United States change back and forth between Daylight Savings Time and Standard Time throughout the year. When the change is made from Standard Time to Daylight Savings Time in the spring, clocks are moved ahead one hour. Within days of the change, most people lose awareness of their following Daylight Savings Time. Trinidad, one time zone to the east of the eastern United States, never changes to Daylight Savings Time. Consequently, during the part of the year in which the United States is on Daylight Savings Time, there is no time difference between Trinidad and the United States. During the time of year when the United States follows Standard Time, Trinidad's clocks are one hour ahead of clocks in the eastern United States.

People also develop expectations about how others will act based on ideas of time. For instance, when one telephones somebody in a different time zone, it is wise to figure out what time it is in that time zone. A Trinidadian calling a relative in Los Angeles at 9 A.M. in Trinidad cannot expect a relative to act as if it were 9 A.M. in Los Angeles. Since the time zones are different, during Standard Time, when it is 9 A.M. in Trinidad, it is 5 A.M. in Los Angeles. When Trinidadians call the United States, then, they try to call at times when their relatives are awake and able to talk— to do this involves using cultural expectations about how somebody acts at a particular time of day and cultural knowledge to calculate what time of day it is where the person being called lives. There are many cases in which Trinidadians contend with ideas of time and expectations of behavior associated with ideas of time. In doing so, they manage situations in which ideas of time and expectations of behavior are distributed across individuals, contexts, and social roles.

### School Time and Taxi Time

In the last chapter I discussed links between occupations and work rhythms. An important component of coordinating different occupations is knowing their associated times and cycles. The case of coordinating taxi drivers and schoolchildren, an important daily occurrence until 1995, demonstrates the importance of reliable expectations of others' time in particular contexts.

There are several taxi drivers who serve Anamat. For the most part, they carry passengers between the village and Viego Grande. They charge fares per person, per trip, and the route and distance traveled determine the rate. The time a trip takes is irrelevant in determining fares. If the road is clear and the traffic is light, it might take as little as twenty-five minutes to make the journey from Anamat to Viego Grande. If construction forces the driver to take lengthy detours, the journey takes longer—as much as two hours. All Anamatian taxis go to Viego Grande and, there, they park at the taxi stand to wait until they fill with passengers, at which time they return to Anamat. Most taxis begin work at the same time each morning in order to capitalize on the large number of people commuting to work, school, or the market, but taxi drivers do not work according to a schedule; doing so is unprofitable, because it results in partially empty trips, and for the car owners, the operating costs of making those trips are the same as when the taxis are full. Furthermore, driving conditions make flexibility an asset for those who drive taxis, and passengers must be willing to tolerate delays from flat tires, accidents, construction, flooding, police roadblocks, and traffic. In effect, scheduling trips is impractical. Still, villagers rely on the predictability of taxis to judge when to leave for work and when they are likely to be able to ride in their favorite taxi. This predictability arises from the rhythm of making trips to and from Anamat, and the rhythm of trips, in turn, is influenced by the condition of the road and when the most people travel. Those who use taxis learn the patterned daily variability of these rhythms.

Those who "work taxi" know these expectations even when not working. The distinction between when a taxi works or not is blurred. Since the taxis are familiar to everyone who travels the road between Anamat and Viego Grande, and residents of Anamat know all the cars, whenever a taxi driver drives his or her car—whether for work or for play—passengers flag down the taxi. Usually, the driver agrees to any request: "Anything to make a shilling," drivers say. Taxi drivers even pick up paying passengers when liming or going to fetes. Temporal flexibility becomes a hallmark of taxi drivers, whether they are working or not. People expect dealings with taxi drivers to involve this sort of time.

For taxi drivers, the rhythms in other people's lives define a large part of the drivers' schedules. Taxi drivers know that there are monthly pay cycles and that when people are paid, the number of people wishing to travel grows.[1] Drivers plan for licensing fees and insurance. For taxi drivers, the rhythms of Trinidadian social life become as the rhythms of the seasons are for farmers. Taxi drivers also recognize the unpredictability and the possibilities of the future. They save money in case they need to repair their car, and they evaluate the possibility of working other routes, which might involve improving their car or buying a new car, or even investing in a "maxi-taxi," a small bus or large van licensed to carry paying passengers. For taxi drivers, then, their work rhythms do not conform to the clock, for reasons embedded in the nature of the work, and by implication, clocks do not determine the temporal expectations others have of taxi drivers.

A very different status is that of students who attend school in Viego Grande. Until the middle of the 1990s, when the government started to subsidize regular transportation for school students to get to school, these students faced the challenge of using taxis to get to school on time. Time in school is very much dictated by the clock, and as long as students had to rely on Anamatian taxis, getting to school was, in many ways, a race against the clock, dependent on the services of taxi drivers whose activities were not linked to the clock.

The experience of time for taxi drivers and students in this context differs, yet they are each aware of the experience of the other. The taxi drivers know the time constraints, defined by the clock, placed on students, and the students know that no matter how urgently they need to get to school on time, the physical constraints of the road limit how quickly a taxi can make a trip back and forth to Viego Grande.

For their part, taxi drivers filled their cars with students when going to Viego Grande in the morning. They would not wait for return fares in Viego Grande but returned, often empty, to Anamat immediately. This reduced the amount of time it took for a round trip. In addition, the students employed strategies that recognized the rhythms of the taxi drivers. Many students would reserve a seat every morning in the taxi of their preference. When ready to go to school, these students would stand by the road waiting for their ride. As the taxi returned to Anamat, it would pass these students as it went up the road to the far end of the village, where the taxis turned around. Upon spotting a student who had reserved a seat, the taxi driver would keep that seat vacant for the student after turning around.

Thus, in this one context, despite different experiences of time, taxi drivers and students coordinate their relationships based on awareness of the others' experience of time. In effect, different concepts of time can coexist in the same context.

## The Distribution of Clock Time

Students contending with clock time forces awareness of the clock on taxi drivers. It turns out that schools play a very important role in the distribution and internalization of the only widespread conception of time in Trinidad—clock time. When in school, children must meet a series of temporal expectations associated with being a student: to be at school on time, to perform activities and to study subjects during times appointed by the teacher, and to leave school on time. Anamatians assume that young children have difficulty in meeting the temporal expectations of being students, but that as they grow older and progress as students, they develop the temporal responsibility associated with their status; that, in turn, influences the evaluation of their development. Schools expect students to be temporally disciplined and attach activity to a rigid schedule with specific activities assigned to carefully defined periods.

The temporal rigidity of school inculcates in children temporal values thought by educators and civil servants to be important in the workplace: awareness of the clock, punctuality, discipline, and single-mindedness in work. This has been a role of European-derived public education since its beginnings (Foucault 1977).

The educational system effectively defines the future for children from the time they enter primary school until at least the time when they can no longer take the Common Entrance Examination to earn a place in secondary school. Once they have earned a place in the secondary school system, children anticipate their exams. The need for the temporal model of school is based upon individuals leaving school and joining the labor force. At this point, the labor trajectories of men and women split. Men often face unemployment and tend to emphasize liming. Women face unemployment as well, but instead of liming, they work in the home.

With regard to time, then, schooling disseminates ideas concerning the clock, temporal discipline, and deference to an institution to define the future. Individuals who go through school learn these ideas and have gained the cultural knowledge necessary to deal with work settings that emphasize these temporal features. Clock time also provides a reference independent of all other social or production rhythms for the coordination of social relationships. In theory, then, the wide distribution of awareness of clock time provides the means to coordinate all social relationships.

### Learning to Lime

Students leave the strict temporal routine found in school after the school day ends. The temporal system associated with students after school inverts school time. After school, children lime and delay going home. They defy schedules and the attachment of single meaningful activities to carefully delimited blocks of time. Thus, on the one hand, children of school age are expected to be temporally disciplined and attach activity to a rigid schedule with specific activities assigned to carefully defined periods and, on the other hand, they are expected to be free of temporal restrictions and the need to make time meaningful. Whereas the school carefully defines the future of students, liming outside the school emphasizes the present. While school effectively trains students to become effective laborers in settings where the clock and temporal discipline are important, the institution of liming, which develops in contrast to the rigid temporal discipline of school, trains students to lack direction and purpose when no institution defines time for them.

The exception to this occurs toward the end of primary school, when students prepare for their Common Entrance Examination. Every year, thousands of Trinidadian students take this exam, and every year the newspapers publish the names of those who received a passing grade. In preparation for the exams, many students spend time studying and reviewing what they learned in primary school, in special Common Entrance Examination preparation classes held after regular school hours. They test themselves by taking practice exams printed in the nationally distributed newspapers. Parents strongly encourage their children to study and, in many households, liming and playing are not permitted unless the student has done some studying. All students go through this trial, but only some students succeed and attend secondary school. The other students begin a transition into the workforce, either by working for their parents or by learning a trade.

## Temporal Disorder[2]

The contrast between the school-inculcated temporal discipline dictated by clocks and the temporal flexibility inherent in liming carries over into adulthood. As discussed in the previous chapter, clocks are not useful to estimate and coordinate activity in much of Anamatian social life. Furthermore, clock time does not define most work available to those who complete their schooling. In agriculture, for instance, work discipline means working until the job is done, rather than working a determined amount of time. Yet the result of the combined effects of schooling and liming is that where institutions do not define time, liming defines time. In effect, the training students get in school does not prepare them for the work world of Anamat in particular, and Trinidad in general.

An example of this problem can be found among the "boys on the block." Most inhabitants of Anamat consider the time in which the block boys live to lack structure and meaning, because this time is not associated with any positively evaluated activity. Most of the boys on the block recently left school with the remainder being young adult men who recently lost their jobs; their ages range from sixteen to about twenty-eight. For the recently retrenched, the block becomes the place to which they return. Villagers interpret the lives of the "boys" as being spent gambling, dealing drugs, consuming drugs, and engaging in various "hustles" to obtain money. According to many, the "block boys" could be engaged in a legitimate way of making money—they could squat on government land and plant ground provisions and bananas, but the block boys choose to "waste" their time. Many do engage in a limited amount of agricultural activity but prefer to await possibilities of wage labor, and they engage in hustles like gambling and drug dealing rather than becoming involved in long-term agricultural projects. To a great extent, they are timeless, in that neither they nor others wish to give their time a social meaning. Their unemployment and underemployment, insofar as it creates a problem for the village, is a problem of what can be given to them to do that would be mutually satisfying for the employer and the employee. Since there is very little employment outside harvest times, the consensual view of their time is that it is wasted.

The time of the block boys is similar to the time of schoolchildren after school. In both cases, time lacks a rigid, regularized structure. In addition, as with the after-school limes, multiple activities occur at the same time among the block boys. When "the block," a couple of walls over culverts along the road, is "open," the block boys play cards, gossip, share information, and engage in a variety of other activities. The time of the block, like the time of children outside school, contradicts the rigid structure and high connection between time and culturally encoded productive activity. The lack of such time in the life on the block creates a social rejection and marginalization of the block, even in the minds of the boys on the block. Their use of the term "boy" implies that those who spend most of their time on the block are in a category with children. Ideologically, by using this metaphor the villagers define those on the block as "boys" and not "men." The time of the block—structureless and meaningless—reflects the lack of social linkages to structured and meaningful times of the boys on the block. In other words, times are made structured and meaningful by the need to organize productive activities in time.

The future, as represented by the boys on the block, consists of the imminent future, except for carnival fetes and gambling tournaments which define the distant future. The boys on the block are young men stripped of the institutional structures of school, with no similar structures to replace school.

For many villagers, school defines time and the future. Without school, time becomes amorphous, and the future is limited to the "forthcoming future" (Bourdieu 1979:8), a future seen as growing out of present conditions. School trains children to be workers with a sense of temporal discipline and a dependence upon an institution to define their future. If an individual is lucky enough to get a job after finishing school, then the employer takes over the role of the school and defines time and the future. For many children, school prevents them from learning other ways of looking at time and other ways of working with time. A common complaint by old heads is that many children have no desire to work in agriculture after they obtain an education. It should be noted that what these young adults want are jobs with the government, not entrepreneurial endeavors. They do not want to start businesses; they want positions in which others carefully lay out their time and their future. Anamatian youngsters see that some people make a very good living off agriculture. Indeed, the wealthiest people in the village are farmers. The boys on the block not only lack the patience for agriculture, but they also lack the ability to define their future. Those who escape this trap are those who were able to engage in some activities other than school and liming when they were children—they worked in the fields, or helped their parents with their business. In effect, school serves to make members of the lower class of Anamat into a potential workforce and deprives them of the skills they need to make an independent living.

### Gender and the Spatial Distribution of Times

In contrast to men, whose evaluation by others is closely linked to remunerative work, women are evaluated in terms of motherhood and work, not necessarily remunerative work. Within Anamat, this is the case regardless of class. This is so throughout the Caribbean, although in the region women also play important economic roles in domestic groups (Henry and Wilson 1975; Lazarus-Black 1994; Safa 1995; Senior 1991). For women, there always seems to be work to do, whether they are paid for it or not. This work includes domestic chores as well as agricultural work. Furthermore, women's liming is not as public as men's. Men lime in rum shops and along the street but not at their workplaces, which makes it easy to spatially distinguish time spent working and time spent liming. Women tend to lime in their homes. This means that there is no spatial distinction between a site of work and where women lime. As a result, temporal distinctions between women's work and women's liming is easily contested.

Men attempt to control women by means of controlling women's time, and women try to subvert this control. Even though women can, and do, perform many of the remunerative activities that men perform, and even though, in some cases, women earn more money than their husbands and male kin, male attitudes remain entrenched in an ideology that defines adult women as having more entitlements than children but fewer than adult men. Men hold women responsible for maintaining the house and children: men expect women to structure their daily rhythms around the demands of the men and children in the house, anticipating their needs. Even so, many women structure their daily routine, whether working on their land or working around the house, to assert autonomy from men.

The most explicit example of this is the practice of watching soap operas on television from noon until 2 P.M. In part, women use this time to escape from household duties. In many households, women's watching soap operas is in defiance of male attempts to control women's time. These television shows intertwine multiple narratives that emphasize gender relations in mythologized North American, usually affluent, settings. The plots tend to center around romances that attempt to overcome repeated obstacles and conspiracies formed in pursuit of social, political, financial, or sexual gain. In talking about these soap operas, Anamatian women tend to focus on the plots that emphasize romantic love overcoming incredible odds, whereas Anamatian men emphasize story lines about villainesses who threaten the downfall of male characters, but whose schemes are ultimately foiled. Some men express concern about women following the examples set by the villainesses in the soap operas. The attempt of these men to prevent women from watching the programs reflects both an attempt to control what women watch on television and an attempt to control when they watch television.

Such was the topic of conversation one afternoon in a taxi filled with men, one of whom was traveling to work. He entered the taxi and, in reference to the soap opera *The Young and the Restless*, he complained, "Now that I gone, I know my wife and daughter goin' to sit down and get young and restless."

Another man responded, "What can you do? That picture come on and cookin' stop, cleanin' stop. . . ."

The first man then interrupted the second: "Only women feelin' they young, and they restless."

Commonly, men believe that not only should a woman manage the domestic affairs of the household but also that she should do it in such a way as to be always coordinated with male schedules and whims, including entertaining the husband's friends. One man said in praise of his wife, "One thing I can always say about my wife: she always entertain. When I have friends over, or if we goin' on a lime, she always cook." In men's views, women should anticipate irregular work schedules and male desires to lime; failure to do so is failure in managing domestic affairs and is thereby failure as a female member of the household. Fulfilling such demands is daunting when men have no consistent work schedule. Those men who own land or squat on land work more with a task orientation than according to a schedule. This makes the time in which such men come home from work somewhat unpredictable. Liming is always unpredictable in its commencement and duration. Women have to rely on numerous strategies to accomplish temporal coordination in the household, and sometimes they do not succeed. Many men overlook occasional problems, although they treat perceived habitual failure in meeting temporal demands in a variety of ways, ranging from public shame to domestic violence. One man told a story about coming home and finding all the pots empty and no food cooked. He said that, in his anger, he complained that with so many women in the house and with so many pots in the house, there should be food in the house. As his anger grew, and nobody began to cook, he began to throw cooking utensils and pots into the street, exclaiming, "Pots are for food. If there's no food in de f——pots, what do we have de f——pots for!" He ended his tale stating that by the time he was about to throw the stove into the street, someone came and began to cook. While there may be a degree of exaggeration in this story, its telling, and the applause and laughter with which other men greeted it, indicates significant shared attitudes about women and the coordination of food preparation with male desires.

Despite such tantrums, and despite threats of domestic violence by some men ("sometimes they [women] get it in their heads to not cook, and yuh mus' hit dem a cuff to set them straight"), many men resign themselves to being unable to get women to do any sort of work during the two hours per day when the soap operas are on television. This subversion of male authority reinforces male concern over women watching soap operas. Men will talk about particular marriages and relationships and complain that women are getting "wrong ideas" from the shows, but the most frequent complaint is that they cannot get their wives, sisters, and daughters to do anything except watch television when the shows are on.

The watching of soap operas for two hours every day poses a serious threat to the accomplishment of domestic synchronization. It is men who ultimately must structure their time to synchronize with women's watching soap operas. In fact, there are some cases where the men must take over the family business while the women in the house watch the shows. Even so, while men protest, they do not usually force women to abstain from watching these shows. They say that this is to "keep the women happy." Men fear angry women. Indeed, if a woman desires, she can make the lives of the men who depend upon her for cooking and cleaning very difficult. In couples, women can refrain from sex with their husbands. If a man's reaction becomes too severe, many women have a large family network to which they can retreat. While men feel that they could have their way and stop women from watching these shows, men suggest that doing so would be too much trouble. Besides, many men enjoy the shows. One gets the sense that men do not want it known that they enjoy the shows as much as women do and that they would rather have the women in the house turn on the television in order to watch the shows.

Viewing soap operas is a subversion of men's control of women's time. Whether or not women emulate the negative behaviors of the soap opera characters—and few, if any, do—the mere fact that they do not submit their time to men's time, and make men submit to women's time for two hours per day, is enough to make some men anxious. The conflict over soap operas is one case of temporal tension between men and women. Its basis lies in the fact that men clearly delineate work space and leisure space, whereas women cannot, because home is where women work and lime.

## Age and the Distribution of Times

Time becomes a major bone of contention between the youths and the old heads. Usually it is older men and women who employ young men. In employment situations, the temporal conception to which one adheres depends upon the employers and the previous situations of employment a person has been in. The men who worked constructing Fort Read during World War II have a very specific conception of time derived from that experience, that they, in turn, utilize in situations where they employ or supervise others. When working on the construction of the base, they had a long commute and worked long hours. They had to be at the base on time and were paid only for the hours they worked. In such a situation, time belonged to the U.S. Army and, on a larger scale, to white colonial powers. The management of this time was in the hands of the officers in charge of the Trinidadian workers. At the same time, the wives of these men were forced to accommodate an intimidating range of tasks. They had to order their lives around the daily rhythms of their husbands. This was one end of a string of domination, in which the men had to accommodate the temporal rigidity of the army, and their wives had to accommodate the men's accommodation of temporal rigidity. In addition, because of the acute food shortage, women were forced to take over any land owned by the household, in order to grow food crops. Likewise, with most of the male workforce obtaining work at high wages with the U.S. military, women played a larger role in working on the plantations. Finally, the government's "Grow More Food Campaign" leased land to landless farmers (see Carrington 1971:124–26; Marsden 1945). Again, the men went to work for the army, and the women stayed in Anamat to cultivate this land for the campaign.

Consequently, wives and mothers became the focal points of a wide range of household activities in which they had to accommodate both their agricultural work and their husbands' and sons' work at Fort Read. From all appearances, the manner in which they accomplished this was, in part, determined by the men enforcing the same sort of temporal rigidity on the women as was enforced on the men at the base. The degree to which this was successful during World War II is difficult to ascertain, but this particular sense of time has been adopted by old heads from that generation, men and women alike, to structure the time of people over whom they now have power. The particular sense of time these workers adopted also influenced their experiences after the war, at the very least providing a rationale for structuring the time of others in a particular way. These men and women, now elderly, hire young adult workers to work on their land. The complaints made by both employee and employer in these situations often reflect disputes about time. The young laborers complain that they are not allowed enough freedom and that they would rather get paid for the work they perform rather than by the hour—complaints reminiscent of those of workers during the Industrial Revolution in England and the United States.[3] They also complain that too much work is expected of them during an hour. The employers, on the other hand, hearken back to their experience on the base, when "men got paid for working rather than liming." Both parties recognize and agree on the behavior of the workers and employers; what they disagree on is the way in which time should be conceived in the work situation. For instance, one man who had worked on the base commented: "In the minds of the younger ones, there is no need for punctuality. If I tell you, 'All right, look, you have to start work here at seven o'clock,' and you come here eight o'clock, I pay you for seven hours instead of eight. I did employ you [for the day], but I paying you by the hour. When the day up, you'll want to get your eight hours' pay. You see where it bring a confusion?"

Research on a similar shift in temporal representations during the Industrial Revolution reveals similar conflicts. American journeymen viewed their work as the completion of tasks (Brody 1989). By implication, they thought that their wages should be determined by what they produced, not by how long they worked. When masters began to estimate how much production could be accomplished in an hour, and how many hours a man could work in a day, and they then calculated the wages for piecework on that basis, the journeymen protested. The journeymen argued that they were not selling their time, but their labor, just as their contemporary rural Trinidadian counterparts now argue.

As a result of such conflict, many employers find it difficult to hire help to work on their land. Whenever time enters the relationship between members of these groups, a dispute erupts. Indeed, both groups anticipate these disputes. The old heads complain that the young are lazy, that they do not want to learn to work, and that they waste their time. The youths complain that the old heads demand too much and that there is plenty of time to do everything.

Often, beginning in their mid-thirties, young adult males who previously associated with the youths, particularly the boys on the block, begin to distance themselves from the block's activities. They begin to squat on government land, using it to grow crops, or they look for sharecropping arrangements with local landowners unable to cultivate all of their land because of a shortage of labor. If these individuals have a skill or some money, they might try to start their own little business, such as buying a car and becoming a taxi driver or becoming a tailor. At this time, such individuals begin to emphasize caring for their family. Indeed, time obtains its value not by association with any particular work but because it is associated with providing money to support one's wife and children.

In many ways, school again defines the future, as the men's children prepare for the Common Entrance Exam. According to these men, time spent on the block is time that could be spent putting food on the table and caring for children. This time still lacks a rigid structure, but it becomes more sequential than that of the boys on the block. No longer do these men arrange their day to watch soap operas, or even to play small-goal soccer. The time of these men is not as rigidly structured as that of school, and, more importantly, the association between time and activity is self-motivated and self-defined, rather than required by an institution, such as school. So there are two causes of the age-influenced distribution of ideas about time: the generational differences, which affect both sexes, and the shift in behavior that occurs mostly in men during their thirties.

### The Power to Determine Time

Some individuals in some statuses, such as bosses and religious leaders, have the power to determine the concepts of time in a particular context. For employers, this power comes from control over wages. Religious leaders can also define time. Local pastors, priests, and pandits are given special power, in this regard, and only charismatic leaders who come from outside the local network of religious specialists are allowed to transcend this power. Indeed, charismatic religious visitors are not only allowed, but expected, to transcend the power of local religious specialists.

Krishdath was an elder in one of the village's Protestant churches. Since the pastor did not live in the district, for most of the week Krishdath was the primary leader in the church. Part of his function was to lead weekly Bible studies. These would be scheduled on a particular night, at a particular time, and at least the young adult members of the congregation were expected to attend. Habitually, these meetings began long after the scheduled time and, during the meeting, Krishdath rambled through an impromptu sermon.

Krishdath's tardiness and his style of preaching defied the time conventions found in both the Hindu and the Catholic practices of giving homilies. Through his use of time, he asserts control over the congregation, and the meeting is spiritually lacking until he arrives and begins. Nothing can begin without him; therefore, the schedule is only as relevant as he chooses to make it. Time becomes sacred because of his presence, and because of this, he has control over time in the context.

His relative Amar, who is the pastor of the church, is very much the same. Time takes on a special value associated with the activity, and not with how long the activity takes. The combined attitudes of these kinsmen toward time becomes acutely contrasted with American missionaries who occasionally visit the church to lead a crusade or a revival meeting. There is one church congregation in the United States that occasionally sends representatives to work at the church in Anamat. The American delegation typically consists of the pastor of the American church and at least one elder and, during their stay, they lead a week-long "crusade." This consists of nightly services aimed at gaining converts and attracting members. In the church, there is a large clock on the rear wall—easily visible to those at the pulpit, but behind the congregation. Under the guidance of the American visitors, the crusade services start promptly at 7 P.M. The beginning of the services consists of prayers and worship songs. Amar and Krishdath are in charge of this portion. Their actions manifest their sense of time and their feeling of the sanctity of church time. For them, God must be appropriately worshipped through prayer and song, and the appropriateness is not determined by the clock. Sometimes, the opening prayer of Krishdath or Amar is as long as the sermon of the guest speaker. The

Americans have a clear notion of a two-hour-long service, however. As a consequence, by the time the worship and prayers are finished, there is typically only about twenty minutes before 9 P.M. The American speaker climbs the stairs to the pulpit, looks at the large clock at the rear of the church, looks at his notes, and then admits that he will need to try to condense the sermon in the time left—a difficult task for men used to giving forty-five-minute sermons in the United States.

Sometimes the time defined by the local religious leader is defied by the charisma of another individual. Charismatic individuals can break the rules and frequently do so. One such case is when an especially charismatic individual is invited to give a sermon during a religious crusade in the village. These crusades are supposed to begin at a specified time and end by a specified time. Within them, the time is structured between singing, sermon, and altar call. Each time period has a group in charge of the associated activities. A charismatic speaker can overrule these leaders, however. In one crusade, the speaker spoke for as long as she liked, defying the structure of the event, and she made use of the entire service. She cut short the period of song and prayer, then she asked the congregation to sing and pray at times during her sermon, and then she encouraged singing and praying for long after the scheduled end of the event, occasionally interrupting with short homilies. Unlike Krishdath and Amar, she was not in charge of this church but was invited from the outside based upon her reputation as a good speaker. Nobody challenged her speaking, praying, or singing out of place, however. There were no polite attempts to end her homilies. She was given free rein. Informants afterwards articulated her unpredictability in presentation as part of her appeal.

The ability to conceive of time in this way is limited to a few spiritual practitioners. In the Pentecostal and Roman Catholic churches, this sort of behavior was allowed only from prominent guest speakers. In fact, one man in the Roman Catholic church also has a tendency to take this view of time during worship, but the priest and the lay ministers always conspire to prematurely (for him) end his preaching. Although he is respected in the community, whenever he tries to exercise authority over the order of worship, or whenever he tries to speak longer than others think is proper, those in charge of the church service, whether priest or lay ministers, find a way to cut short his discussions. The charisma involved in defying temporal expectations of religious leaders, in part, consists of the special status of being an outsider and, as a consequence, the expectations that conventional temporal expectations would be defied.

## Conclusion

The purpose of this chapter has been to explore the manner in which temporal expectations are distributed in Anamat. With regard to some contexts, it is clear that there are different associated times that must be coordinated; such is the case with taxi drivers and children. In still other cases, the distribution of times results in conflict, particularly in the case of gender relations and age relations. Finally, sometimes, the associated times are not as clear, such as with religious leaders. In fact, to attach temporal models to particular positions in society might be unwarranted, although this is exactly what Anamatians do. In effect, Anamatians act in a world where times are multiple and distributed. The way in which these models are actually distributed may differ from the assumptions of distribution that cultural actors make.

# Institutional and Consensual
## 4 Temporal Coordination

Within Trinidad, many institutions have explicit rules of conduct in re-lation to time. These include schools, factories, and offices. Those with power in such institutions guard the institutions' temporal order; the plantation overseer, for instance, "keeps the time" for the work gangs by sounding the plantation horn or bell. Indeed, a mark of power in these settings is to be able to define time or to enforce temporal definitions. Leaders in institutions accomplish this only if two conditions are met: a clear framing, often both spatially and temporally, of the institution, and a recognized, legitimate status that allows them to define time and apply sanctions.

Institutionalized ideas concerning time provide guides for coordinat-ing people and activities by forcing common times and expectations. The context of school, as I explained in the last chapter, is an example of where time is carefully defined, in this case by the clock, and rules concerning the links between activities and periods of time are specified. Such insti-tutionalized definitions need not be based solely on the clock, however. In some cases, the institutionalized definitions involve conflicts between the clock and other cycles of activity.

## Cricket and the Clock

The game of cricket involves institutionalized rhythms existing in tension with clock time. As C. L. R. James demonstrates in his book *Beyond a Boundary*, cricket is more than a competition: it also represents many of the tensions of British colonialism (1993). This includes temporal tensions. During the era of the emergence of West Indian cricket in the nineteenth century, the British colonial system emphasized punctuality as defined by the clock. Embedded in the game are the demonstrations of all sorts of ethical and aesthetic codes concerning starting, ending, and breaking for lunch and for tea time. Players and fans express ideas about the pace of the game in relationship to time. While a cricket match can be punctuated by the clock for purposes of a lunch break or tea time, the players actually dictate the rhythms of the game. These rhythms often leave the calling of lunch or tea time up to the judgment of the umpire, rather than in strict accordance with a schedule. Nature also takes its toll—a rain storm may prompt an early break, for instance. In many respects, cricket is similar to baseball in its relationship to clock time. Baseball challenges clock time with its rhythms and cycles that defy minutes and hours (Shore 1996:77). As is the case with baseball, cricket is accused of being a slow-paced game, but just as with baseball, cricket fans do not deny the pace of the game. According to baseball's fans, part of the sport's charm is its irregular pace, with its lulls in action occasionally broken by flurries of activity (Shore 1996:77). Cricket fans make similar comments.

Cricket is played on a field that can be irregular in size and shape but which is traditionally some form of ellipse. In the center of the field, or "ground," is a "pitch"—a narrow, long, rectangular area in which the two batters, or "batsmen," run. At both ends of the pitch are wickets. A wicket consists of three stakes, on top of which are balanced two pieces of wood called "bails." At each wicket stands a batsman whose job is to hit the ball to score runs and to prevent the bails from being knocked off the wicket by the ball. As long as the bails do not fall, and a ball the batsman hits is not caught, or the batsman does not use his leg to prevent the ball from hitting the wicket, the batsman continues to bat. To score runs, the batsman can hit the ball anywhere—there is no foul territory. A ball hit out of the ground in the air is worth six runs, a ball hit out of the ground on the ground is worth four runs, and any time a ball is hit and the two batsmen can run the length of the pitch and cross is worth one run.

The game is played in "overs" and "innings." Each over consists of six bowls. A bowl is when a bowler standing behind one wicket runs and then throws the ball down the pitch in the direction of the batsman at the other wicket. Bowlers can use a wide variety of techniques either to get the ball past the batsman to knock down the bails or to get the batsman to hit the ball in the air for a fielder to catch it. Through control over the spin of the ball, bowlers can alter the flight of the ball, as well as the behavior of the ball after it bounces. Some bowlers focus on developing different spins at slow velocities, and other bowlers use high velocity to attempt to get the batsmen out. The only formal restrictions on bowling are that the bowler must release the ball by a specified point on the pitch, and that he may not bend his arm when throwing the ball. Each day of a test, the umpires try to "fit in" a set number of overs before evening falls and daylight is too dim to play. An inning consists of ten outs, so that when one team has accumulated ten outs, the other team comes to bat.

Cricket involves a great deal of strategy that balances scoring, bowler-batsman match-ups, and controlling time. A good captain will take into account weather and light conditions to make judgments. For instance, if a team comes to bat late in the day, when the light is failing, the captain may choose to put in "night watchmen," that is, good defensive batsmen who are sometimes not the star batsmen on the team. These night watchmen minimize the risk of an out (a "wicket falling"). Cricket matches usually have a set amount of time, ranging from an afternoon to several days. The more of that time one's team is at bat, the less opportunity the opponent has to score. This can motivate some teams to opt to "defend the wicket," rather than to emphasize scoring. In this case, the batsmen primarily attempt to make contact with the ball to prevent it from hitting the wicket. An aggressive, offensive style is often riskier, in that such a style does not allow for as strong a defense of the wicket. An added dimension is the rate at which the bowler bowls, because this determines how much time an over takes. Thus, within the natural confines of daylight, there are other rhythms of "at bats" and "overs" that influence the game.

If the time allotted for the game expires before both sides have finished their at bats, then the game is declared a draw, regardless of who had the lead when the game ended. It is possible for a team with inferior talent to play to a draw with a superior team by playing defensively, although such play is viewed as aesthetically repulsive. As a result, one element of strategy in cricket is balancing the amount of time one wants the innings to take with the amount of run production the team needs per over, and the amount of time before sunset or when the game is over. The production of batters is evaluated in terms of their total number of runs, and the number of runs they score per over. The production of bowlers is based on the number of overs they bowl related to the number of batters they get out and the number of runs they allow per over.

Thus, cricket manifests its own time, largely defined by overs and sunlight. It is very important for the competing teams to manage this time. Clock-determined lunch and tea breaks disrupt the rhythms of play and sometimes disrupt one team's momentum by coming at critical times. How each team utilizes their overs can determine the margin between victory, a draw, and defeat.

The Anamat cricket team plays in a local league. Every weekend during the local cricket season (around March through May), the team plays. Usually, the match starts in the afternoon, but the starting time is dictated by when all the members of both teams arrive and when an umpire can be found. Once the game begins, it proceeds at its own pace until the umpire calls a break or until the umpire determines that it is too dark to play. In effect, while there are clear rhythms and time constraints embedded within the confines of the game, the game itself is embedded within multiple temporal constraints of individual lives: work, seasonal rhythms, and social relationships.

## Legitimacy and Time

Whether a leader is a teacher, a foreman, or a cricket captain, getting people to follow a definition of time depends on their recognition of one's power and legitimacy to define time. People defy institutionally defined times when legitimacy is lacking. While Roman Catholic mass starts promptly at 7:30 every Sunday morning, when Anamatian Roman Catholics gather outside the context of the mass to work on the church or to meet to make plans for the church, temporal conflicts arise. Some, who work in the fields all day, come to the meeting after returning home, bathing, eating, and changing clothes. As a result of daily fluctuations in work habits, sometimes they arrive early, sometimes late. Others, whose lives follow a clock-determined schedule, arrive punctually and complain about the tardiness of others. The basis for their complaining is that, while the meeting might be set for six o'clock, people still arrive at seven and even seven-thirty.

Event participants often deny event organizers the legitimacy that would be necessary for them to define time. "Excursions" are an example of futile attempts to define time, due to leadership that is powerless to enforce temporal standards. Excursions are organized trips that require the purchasing of tickets to participate. They take the form of mobile fetes. The organizers charter a maxi-taxi (a van or small bus) to take the ticket-holders to a popular destination, often a distant beach. Organizers say that, because of the distance involved and the cost of chartering the maxi-taxi, excursions must leave at a scheduled time early in the morning. Excursions never leave on time, however: "They tell you six o'clock, bus leaving at six o'clock. The bus might reach, the bus might reach about ten to six in the morning. And the bus taking about forty. Twenty might come and the balance have to wait."

Those who go on excursions view these events as similar to fetes. In fetes, nobody has authority or power over the others, and such events begin when all who have tickets arrive and end when they want to go home—"You limin,' not workin,'" one man quipped. The organizers have little authority, and if the organizers ordered the maxi-taxi to leave on time, thereby leaving behind tardy ticket-holders, the organizers would lose legitimacy to sponsor excursions in the future.

Thus, institutions can contain within them definitions of time, but effectiveness of such definitions depends on someone within the institution possessing legitimacy to use power to enforce them. Even then, leaders of institutions that attempt to define time recognize, sometimes resignedly, that their institution is only one part of the lives of their participants, most of whom manage conflicting temporal expectations.

The context of Hindu rituals in Anamat provides an example of legitimacy. Hindu rituals are held at particular, auspicious times. In the case of the Hindu pujahs held every month in Anamat, the ceremony takes place at twilight, because the period between day and night is an auspicious time in Hinduism. Twilight is defined as 6 P.M., and since Trinidad is in the tropics, the actual variation of when the sun sets during the year is not very great. This definition of time binds both devotees and pandits. Since all the pandits who serve Anamat live outside the village, they wish to have the event ready to begin when they arrive. Their sacred function gives them legitimacy to define time.

Within the Roman Catholic Church, the sacred authority of the priest and his representatives provides legitimacy to their definition of time. The priest has several masses to officiate every Sunday morning and must be punctual in the beginning and ending of the mass in each village, so as to make it to mass in the next village. Therefore, the priest ensures that mass begins on time and ends on time. Everybody is aware of what time mass is held, and that the priest need not wait for latecomers before beginning the sacrament.

For both Catholics and Hindus, the time of the religious ceremony is sacred—it is a divine possession. Both pandits and priests argue that tardiness to services signifies a lack of respect for God as well as for God's representative in the form of the priest, pastor, or pandit. As a result, most who participate in these services appear at the time set by their religious leader. Then, they defer to the leader's power to control time. While services start punctually, they might not end punctually, depending upon the aims of the priest, pastor, or pandit. In the Roman Catholic Church, special services often extend several hours longer than normal services, but this is done under the priest's leadership. Pujas are determined by the completion of the various phases of the ritual, as well as by the explanation of the puja performed by the officiating pandits. This explanation often takes the form of explaining what was performed in the puja and a homily on how a proper Hindu should behave.

People learn the time of religious ceremonies by growing up participating in these contexts. The burden of preparing the mandir (Hindu temple) or church and starting the services often falls upon adolescents, who must be prepared to begin before the arrival of the pandit or priest. The youth group sweeps the floor, cleans the building, and then forms a singing group to provide music for the service. In all of the religious groups with singing groups in Anamat, the oldest adolescents organize the choirs with the help of some young adults. This is a risky endeavor for the adolescents: they recognize the heightened temporal responsibility placed upon them and realize that when they fail to meet their responsibility they can be publicly humiliated, a punishment beyond the indirect verbal chastising that adults might receive for a similar oversight.

Often, the oldest members of these groups take charge and chasten the younger members. One Sunday morning on which I was to accompany the Roman Catholic youth group on guitar, I arrived before the service to find the church a bustle of activity. Several teenage girls were supervising the preparation of the sanctuary for mass, because "Father [the priest] was coming." In the midst of the youths hastening to sweep, clean, and organize the pews, one of the oldest girls there said loudly, "Oh gorsh, you want to make Miss Cynthia [one of the lay ministers] vex? Father comin' jus' now, and the floor ain't even swept!" Hindu youth might be barred from participation in *pujas*, which effectively excludes them from a social network of young Hindu peers. Since there is an ideal of village exogamy and parents limit contact between young men and young women, the ritual settings in which Hindu youth groups participate are important for meeting and befriending members of the opposite sex who are potential mates (Freilich 1991). The Roman Catholic youth group feels the threat of sanctions even more strongly than the Hindu youth group does, because the Roman Catholic youths are usually members of a confirmation class, and their ability to manage responsibilities given them within the church is evaluated by the priest and the lay ministers in determining who will be confirmed.

### Coercion and Time

It is not the case that all formal institutions rely simply upon legitimacy to establish temporal definitions. Where legitimacy does not exist, coercion is attempted. For instance, many employers use coercion. Tangible symbols of the definition of time, such as a time clock or the position of the sun, also provide the means to propagate and enforce the appropriate temporal standard. When individuals are hired, they are told when to appear at work, how they will get paid, and penalties for "idleness." This varies from employer to employer.

Factories have been seen as institutions that introduce a particular type of time found only in industrial capitalism—a time necessary for industrial capitalist production (Thompson 1967). Time obtains an exchange value by which labor is measured (Marx 1977:683–91). Since time is a means of determining labor's exchange value, even task work in capitalism "is nothing but a converted form of the time wage" (Marx 1977:692). Both the laborer and the employer wish to measure time carefully and to maintain some standards of amounts of work per time period. In recent years in Trinidad, factory workers have been punching time cards in time clocks before and after work. Employers schedule breaks at certain times, and workers must work a certain number of hours per day; and, because employees punch time cards in a time clock, the resulting record is supposed to show employers if any particular worker arrived late for work or left the job early. This rigid structuring of time resembles that of school, although the punishment for tardiness at work is having money subtracted from one's paycheck rather than the corporal punishment found in school.

As discussed in chapter 1, Anamat's plantations followed a form of time similar to that of factories. They had a strict regimen punctuated by the blowing of a horn or the ringing of a bell. Plantations gave one signal ten minutes before the beginning of the workday and a second signal when the workday started. The overseer had the authority to tell any worker who appeared after the second signal that he must return home without work or pay. The overseer assigned those who appeared on time to work in gangs, which he then deployed in the fields. The bell or horn also marked break times, lunch, and the end of the day. On Anamatian plantations, overseers used gangs to manage the labor force and bells or horns to temporally structure the day both when paying workers by the task or by the day.

Since plantations were highly centralized, and workers reported to the overseer before and after work, workers had difficulty challenging the plantation's temporal regimen. In addition, the cocoa plantations in Anamat were not very large—much smaller than sugar plantations. This relatively small size made it possible for overseers to visit every work gang and carefully supervise many tasks.

Before Trinidad and Tobago's independence in 1962, the Department of Public Works[1] imposed a system of time on work conceptually similar to that found on plantations except that the Department of Public Works relied much more heavily on delegating time-keeping authority to foremen, who were required to carry watches. The Department of Public

Works did this because the roads in many districts, including Anamat, stretched for many miles. Therefore, unlike the plantations, the workers were not always within range of a centrally controlled signal, such as a horn or a bell. As it turned out, when the supervisor and overseers relied on foremen to keep time, the foremen incorporated temporal flexibility into their scheduling on the work site.

During the 1950s, the engineer, the official in charge of the local Department of Public Works, and some of the overseers were British, rather than Trinidadian. The expatriates distrusted the local foremen's desire, and even ability, to keep time. They expressed their distrust by policing the work gangs around lunch and quitting time. One man, who worked for the Department of Public Works for many years, and who has since retired, described how his superiors carefully monitored him. He eventually "proved" his temporal trustworthiness, but it was a long process. When he began working for the Works Department, he had to watch the time carefully:

> Anywhere that I working, I cannot leave before half past four. Even the superintendent would pass where I working, so I had to remain there—reach there in time and leave there on time. One day I was [in] Granital, I remember, the superintendent was passing—that was after three in the evening. [He] turn back and watch again, but I see him, but I do as if I didn't see him . . .
>
> Another day, I meet him [while I was working on] Belagua Road. Between here and Belagua Road now, he stop. He say, "How they call this road?"
>
> I say, "This is Belagua Road, sir."
>
> He say, "Oh." But it was not that [which he came to ask me]. He pass to see if I was working, or something or the other. I had the supervisor that is under him. One day I was working on the Tomacito Road. He pass there about two o'clock. He stop, he speak, and everything else, he left . . . Well, that bridge had finished. I ride a little way to a next bridge. It had a road junction they called Brahma Road. I didn't know that this man came there and parked his van there. Well, I came and start cleaning up the bridge and thing because I have something to do; and that time, it was five minutes to four . . . I hear the van start out inside there,

and he drive out to me. He came and started talking. He say, "How you going?"

I say, "I going all right."

He say, "By the way, what is the time now?" Well, I always have my timepiece with me, so I tell him it is five to four. He say, "Oh it is too late; it is too late. I have to go to Neelim to the office, but I wouldn't reach in time." But it was not that—it was me he was tracing.

Even with such surveillance, workers deliberately attempted to defy these policing attempts and lengthen breaks or shorten the workday. Local foremen developed skills to avoid sanctions for not adhering to the prescribed daily schedules. The most common strategy involved adjusting one's watch. A retired foreman described to me a trick of the trade:

We used to have fob in our pants [in] those days. A pocket here, between the waistband here, and you put your watch or your money there. [He points to his waistband and demonstrates where this small pocket was attached.] Well, as foreman, you must have a watch, and I was so versed at that [adjusting his watch], if I want to move the time forward or backward a five or ten minutes with my fingers . . . I put my hand [into my pocket] to take out my watch. By the time it come out, that time is not there again. If I want it back, it is back. If I want it forward, it is forward.

He then described an instance when a foreman he worked for used this trick:

There was a supervisor who had the habit of appearing five minutes before quitting time. One day, the foreman had ended work early, and all the workers were walking home when the supervisor passed. The supervisor looked at his watch and said, "What's the time?"

The foreman reached for his pocket watch, and while pulling it out of his pocket, he adjusted the time. He then said, "Some minutes to five," and showed the watch to the supervisor.

The supervisor looked at the foreman's watch, saying, "My time is twenty-five past four."

The foreman shrugged, and held up his watch, replying, "This is the time I work with."

The supervisor, realizing that he could do little more than verbally chastise the foreman, said, "I want to point out to you that you have twelve men, and five minutes work for your gang is one hour's work for one man, and you know how much work can be done by one man in an hour—a lot."

Such challenges to the official definition of time, when caught, were met with punitive measures. On occasion, a foreman would have his work papers open to record who was showing up for work, even if they were late. Normally, the sheet was expected to be put away at starting time, and whoever showed up late was sent home. Many foremen practiced this temporal deception, however; occasionally a supervisor would appear in the first fifteen minutes of the workday and see the sheet still open, and the foreman would be fired for this offense. Any time a supervisor could verify some sort of violation, the individual would be fired.

After independence, with the growing power of unions and the disappearance of expatriate management, management practices changed. The ability of overseers to punish employees who appeared late decreased because unions contested their authority to do so. The ability to enforce time in public sector workplaces weakened. Whereas previously, the overseer had the discretion of sending a tardy worker home, now the unions challenged such action. Whatever the actual intention of policy makers in making these changes, they have had the effect of increasing the temporal flexibility of government work. This has allowed workers to more easily manage multiple occupations, such as coordinating various forms of agriculture with wage labor. Increased flexibility has led to a view that the relationship between being paid for a day's worth of labor and the amount of time it takes to complete a day's worth of labor is negotiable. Robert, a landowner who also works on the roads, discussed this flexible relationship between time worked and work performed:

> Robert: Now I always use this expression, "Hours don't make work." And I told a big boss that—my superior, an officer—[one] day. He talk about what is the working hours, [starting ] at seven o'clock. I say, "I [am] a working man; a working man never wait

on time." A real working man who like to do his work and thing, a man who has the spirit to work, the will to work, he never really look at time . . . If you work for the whole day, the foreman need to work seven to eleven, twelve to four on the sheet, but in task, the only time is seven, and he inspect our task to see it's complete. If it's half past seven, and you finish your task, and you do the quantity of work, he has no quarrel with you because you do the amount of work. That is the important thing.

K.B.: You said, "Hours don't make work." What do you mean by that?

Robert: Is the amount the man can put down under the same condition. It is how you compare the work . . . It [is] the work the man produce, and the ability.

So, even though employers often equate hours worked with labor performed, this individual and many others argue that the work performed cannot be measured by a clock.

During the 1970s and early 1980s, many men in Anamat worked in the Amalgamated automobile assembly plant several miles away. They worked one of two shifts. At the beginning and end of each shift, workers would punch their time cards in the time clock. If one was late, then one's pay was deducted for each quarter hour of tardiness. The time clock symbolized the link between work and time—a relationship strictly enforced by management through the use of this machine that monitored workers' temporal habits. By doing so, management hoped to have all employees follow the same definition of time: a definition in which the relationship between production, wages, and time was carefully thought out.

Many of the workers were also small-scale, independent farmers, an occupation that involves very different rhythms from factory work, because temporal pressures in such farming come from nature-driven and market-driven agricultural cycles, not from a supervisor or foreman. If they worked second shift, which began at 2 P.M., they might go into the field in the morning to work their land, then return home, change, and go to work at the factory. Sometimes they would arrive at the factory late. This had three possible outcomes. First, a few would just accept the loss in pay. Second, some would have friends punch their time cards, so that the records would not show their late arrival. Third, a few would rather not work at all than arrive late, and they would just take a sick day, return home, and work in the field.

The first and largest group accepted the same definition of time as employed by the factory's management. They viewed time as a divisible commodity: one sold one's labor in the form of time worked. One former employee said, "You come to work for an honest day's pay and you know you come to work for that day, you come to work eight hours . . . I come to perform an honest day's work." Arriving late, one did not work for that period of time and should not get paid for it.

The second group viewed work along the lines of tasks. They felt that, even when one gets an hourly wage, one is being paid to complete a standard number of tasks in the day, and the day is actually defined by that number of tasks. One man said, "Bosses only take action when work not done. Bosses try to make workers 'fraid of them, but they [the bosses] only check your time if you are a poor worker—if you a good worker, the bosses won't check your time very often." In some cases, workers feel they know their work better than their supervisors, and from their point of view, if management does not define what the expected amount of work is, then it is up to the worker to decide on a reasonable amount of work to be completed. According to these workers, arriving late should not make a difference, and being penalized for arriving late is simply management trying to exploit the workers.

The third group saw labor as being sold in the form of a day's work. If one arrives late, then one is not able to work for an entire day. This is probably a remnant of plantation influence, since plantation overseers sent late workers home without work or pay. Contemporary factories pay by the hour, and a tardy worker's pay is simply reduced as a penalty for lateness. Habitual tardiness results in comments in a personnel file and possible suspension or firing. Even understanding the factories' penalty for being late to work, some workers would rather return home, losing an entire day's wages. In doing so, they take a day of sick leave. A former worker at the automobile plant recalled that the factory allowed employees fifteen minutes' grace before they started to reduce their pay, and he added:

If one morning I was to reach there after that ten or fifteen minutes' grace, I would come straight back home. Yeah, I would come straight back home, just to avoid being late. My foreman used to quarrel with me. He say, "You here already."

I say, "Well, I'm late."

He say I reach quite [to the factory] and I went back home.

I said, "yeah, I was late, that is how I operate."

As the automobile plant's imposition of its definition of time reveals, even attempts to make everyone in a particular context follow the same definitions of time bring about a wide variety of reactions and unintended interpretations. For this reason, as a strategy to avoid temporal conflicts, a formal definition of time cannot be completely successful. Still, such definitions delay and limit temporal conflicts. In the case of those workers who would rather not work than arrive at work late, eventually a conflict arose between them and their supervisors, in which the supervisors would castigate the workers for taking too many "sick days." In dealing with employees, the factory's management did not acknowledge the difference in their workers' concepts of time or the different interpretations workers made of the normative time concepts of the factory. Still, the shifts were relatively coordinated and allowed the factory to function reasonably well, albeit never efficiently, in the eyes of those who ran the factory.

### Clock Time and Power

As a result of Trinidad's colonial heritage, and the influence of multinational corporations with predominantly European or North American managers, Trinidadians associate particular ideas of time with particular ethnic or racial categories. This affects legitimacy of an individual to define time influencing the ability of a supervisor to enforce a particular temporal concept. Anamatian workers evaluated management at the automobile plant as extremely rigid when compared to foremen and supervisors in the government ministries. In part, this is because the factory's managers were white Americans and Europeans, both of whom are groups thought by Trinidadians to have different attitudes toward time than those held by Trinidadians. Before independence in 1962, many of the supervisors in the government ministries were also expatriates, and Anamatians applied the same view to these foreigners. Workers feel that they are less able to exercise a flexible conception of time with white expatriate supervisors because such supervisors enforce rigid conceptions of time. According to Anamatians, if the supervisor is Indian or Creole, the situation is different, even if the context suggests that the same rigid, clock time applies. As an elderly man, who had worked on the American military base during World War II and for the Public Works Department afterward, said: "The people from [outside Trinidad], [were] English people and maybe American, different people. So now the Black people, as we may call them, was more obedient to these [English and American] people than to the Black man himself. You find if a white man [is in charge], he's the boss from the time he reach. [Now that white men are no longer in charge], not so again." Many Anamatians who had worked for white Europeans and Americans said that they were more likely to show up late with Creole and Indian supervisors than with white supervisors. These individuals also added that they would be more likely to use the phrase "Any time is Trinidad time" as an excuse to Creole and Indian supervisors than to white supervisors. The degree to which supervisors accept this view of Trinidadian time as a legitimate excuse varies according to individual, and not according to race or ethnicity, however. Both Creole and Indian supervisors express frustration at this temporal attitude but feel that they have been able to do less and less about it.

In some cases, the conflict between enforcers of a definition of time and those having the definition thrust upon them is blatant, conscious, and understood by all involved. Such a case was the enforcement of a curfew following the attempted coup in late July 1990. Law enforcement officers had authority under the state of emergency to shoot on sight anyone who broke the curfew. Despite this, some young adult men refused to obey the curfew and remained liming on the street during the hours when the curfew was enforced. As one commented, "They say, 'Curfew,' and we say, 'Any time is Trinidad time.' They say, 'Six o'clock you supposed to be inside.' You might meet a fellow outside at nine o'clock. You say, 'You supposed to be inside—curfew time is six o'clock.' [They reply], 'Time is time, man—any time is time for me,' and then they go on their way." These men, most of whom had no steady work, complained that the curfew restricted their freedom and that it was against the nature of Trinidadians to stay in their houses for that long period every day. The premium they placed on temporal freedom was met with occasional coercion on the part of police making a trip to Anamat. At these times it was quite clear that the police were unwilling to let curfew breakers go without punishment, even in a remote village far from the crisis. While the police had permission to shoot on sight anyone who violated the curfew, in Anamat they never utilized it. Indeed, most curfew breakers were simply warned, a few were frisked, and only a couple were punished by being beaten, although not seriously. Many of those who remembered the curfew associated with the mutiny in the army in 1970 commented that even this was more coercive than the police's actions in 1970. They said that, during the crisis in 1970, people commonly limed along the street during curfew hours, and the police would merely pass and wave.

The young adult men were not the only ones to defy the curfew. During the state of emergency, a prominent Hindu Indian in the village died. Everyone wondered whether they would hold a wake at night, as is traditional, even though curfew began at 6 P.M. The family held a wake, and it continued long past curfew. Men, women, and children of all ethnic groups in Anamat attended, defying the curfew. Those who attended knew that if they met police while going home, they might be arrested or beaten. All recognized that holding the wake was an open defiance of the regulations of the state of emergency, which also forbade the gathering of large groups of people. Still, Anamatians agreed that holding a wake for this individual was more important than the state of emergency. The death of a prominent resident required the performance of the appropriate traditions, despite the government's curfew. Time as defined by the death of a prominent member of the community was more important than the government-defined time.

In some cases, individuals and groups resist formal institutions' attempts to define and control time. The imposition of time discipline within the factory is met with varied responses, ranging from the assertion that it contradicts Trinidadian character, to simply misinterpreting the temporal discipline embedded within the factory. In the case of the wake, the emphasis on tradition led to the disregard of government rules and regulations. The multiple ideas of time and their links to multiple issues, such as work or government power, result in temporal conflicts, and while the conflict might be temporarily ignored, when the choice between the time of the individual in Anamat and the time of the government or factory is forced upon Anamatians, they inevitably choose their own definitions of time.

Thus, institutionalized enforcement of time has limited success, and it results in tension and conflict. The association of time and expectations forms the basis of punishment when expectations are not met. The more informal the legitimacy and the less serious the threat of coercion, the less the definition of time is respected, as in the case of excursions. Where punishment is serious and enforcement swift, as on plantations or among the Roman Catholic youths, the institutionalized definitions of time are adopted. In cases such as the factory, the workers and the management do not follow the same definitions of time but do not realize their differences. In fact, both groups think they understand one another, but many workers interpret time in terms of other work situations, particularly agricultural situations. The managers base their temporal ideas on a relationship between work, production, wages, and time, but the workers do not. Finally, the more resources that are available to circumvent an enforced temporal model, the more likely the model will lose force. In the case of the curfew, even though the consequences of being apprehended by the police were potentially severe, the likelihood of being apprehended decreased, because of the available resource of endless cocoa fields into which to disappear if one heard a car engine on the road after curfew. The boundaries of these institutions are clearly set, and outside the boundaries, the formal definition of time dissipates.

## Universal Time

Trinidadians sometimes refer to clocks to schedule activities. This differs from an institutionalized definition of time, in which positive and negative sanctions enforce the temporal definitions. In many cases of interpersonal relations, the use of clocks is consensual. Using a widely shared idea of time based on consensus rather than on authority is an example of using what I call a "universal time." Such ideas of time are universal, in only a loose sense in that knowledge about them is widely shared. Such knowledge includes concepts of day versus night, or ideas of hours, minutes, and seconds. This sharing allows people to make common references to universal times.

The use of a universal time assumes that everyone involved in the interaction is familiar with it and is willing to structure their lives around the universal time. The seemingly simple task of scheduling an appointment at, say, two o'clock involves the assumption of familiarity with clock time and shared ideas about its legitimacy. With regard to clock time in Trinidad, it is not the case that everyone can read a clock. Those who cannot use clocks still understand what two o'clock is, but they rely upon people who can read clocks to know the time, or they associate a particular hour with a time of day—for example, morning, afternoon, evening. In addition, many who can tell time do not in any way structure their lives around the passing of hours or minutes. They see clock time as a reference for parts of the day, rather than something to determine when they do a particular activity. Finally, some both understand and apply clock time. When they make an appointment, they will be there at the appointed time, and they expect others to be, too.

With these differences in individuals' relationship to clock time, use of this universal time as a means to avoid temporal conflict is not very effective, because different values are placed upon clock time. Even where there is a shared ability to determine the time, there is not a shared value placed upon the time. A doctor might make an appointment with a patient for one o'clock, and the patient will appear much later or much earlier, explaining that being punctual is secondary to appearing.

Some people, such as doctors or anthropologists, are frequently associated with using clock time as a universal time to arrange their social routine. While Trinidadians recognize that the individual utilizes clock time, many will only loosely follow any reference made to clock time to arrange a meeting.

Clock time is one example of a universal time. There are universal times other than clock-determined time. The time of the day—for example, afternoon or evening—also serves as a universal time, albeit not as exact as clock time. Trinidadians divide the day into morning, afternoon, evening, and night. These only roughly map onto clock time, and it is not unusual to be told, "Good morning" at 1 P.M. or to have one person say, "Good afternoon" at the same time another says, "Good evening." These time periods are indexed by circadian rhythms and the sun rather than by watches and clocks.

Often, Trinidadians use universal times to calibrate other definitions of time. An example of this often arose when I tried to schedule interviews. For instance, I had scheduled an interview with one individual for around 6 P.M. When I arrived, he told me that he was wondering if I was going to show up or not, because he had come home from working his land, bathed, and eaten, and the boys on the block had been playing soccer for some time—an activity that they begin in the evening. Furthermore, this man explained, it was "gettin' towards dark," which meant it was getting close to "night," and an appointment for 6 P.M. is considered an appointment for the "evening."

Conceptions of weeks, months, and years also serve as universally recognized temporal systems. While historical events can be plotted in terms of years, and future events scheduled according to dates, many activities and events are not reckoned according to these standards. This became evident in informants' discussion of the phrase "long time." This idiom is used to refer to something that happened a long time ago, but the exact meaning is negotiated in conversation. During one interview, a man referred to a concrete roller used to maintain the cricket ground in the community:

> I will tell you long time they build that [roller]. If you want to know, you have to ask me questions and then you get to know how long they build that roller there. [It is] a good roller for the [cricket] ground. Long time that there. [You ask], "What long time mean?" Right? And then I will tell you approximately thirty years, or what I mean by that "long time" towards such that I was speaking about. Then I will explain, but before I explain, I jus' tell you, "Long time."

In this case, the memory of building the roller, that had occurred "long time," is related to a number of years. In contrast, on the opposite side of the cricket ground lives a man who said that twenty-five or thirty years was not "long time," because the memories from those periods were "plain" and "not fading." He recalls the start of World War II as "yesterday." He points out that the war ended in 1945 but then adds about the events of the time: "Those were things, memories clear, clear, clear, and everything that most people could remember. Though, when you look at it, let's say at the end of '45 to '90, that would be forty-five years, and yet I don't consider that long."

In both cases, these men related "long time" to when they were young, and their use of the phrase differed. Yet, both related "long time" to a universal time by using either dates or a specified number of years.

To a great extent, the use of a universal time is a strategy of last resort in relationships that require some temporal coordination but that have no formal institution or particular individual to enforce and orchestrate the coordination. The use of universal time has embedded within it an expectation that others know the time and will respond appropriately. Similarly, Trinidadians who must rely on this strategy recognize its problems and have at hand many corrective strategies for when the temporal expectations based upon the universal time are not met. These strategies will be discussed in the next chapter.

### Parang

There are other contexts with clear boundaries, but assumptions about time that stand in explicit contrast to the institutionalized times found in, say, plantations and factories. One such context is *parang*. *Parang* refers both to a traditional Venezuelan-derived Christmas music and to the practice of traveling from house to house to perform the music. *Parang* songs consist of melodies and chord progressions to which verses and choruses that commemorate the Christmas story are set. Usually a lead singer sings a verse and the rest of the *parang* band sings the chorus. The traditional performance of *parang* plays with time: both clock time and the daily cycle of night and day.

During the Christmas season, as soon as night falls, *paranderos* (*parang* singers and musicians) begin talking about "organizin' a paran'." While ostensibly informal, many bands have core members who have played together every Christmas for decades. During Christmas, two or three *paranderos* start visiting the homes of other *paranderos*. Such visitation is a call to *parang* with the band. The new member shares drinks with the band, and then the band moves on. Eventually, when the instrumentation is complete, the band begins visiting homes of friends and relatives.

As the lights go out and people go to bed, the *parang* band begins to move stealthily through the village. They choose a target and attempt to climb the stairs into the front gallery of the home as quietly as can be done with their percussion and stringed instruments. Once the singers and instrumentalists are in place, the band breaks out into song, waking those who are sleeping. The song continues, and gradually lights are turned on, and people inside the house hurriedly get out their stocks of rum, beer, soda, wine, and any other beverages on hand. Eventually, a sleepy-looking but smiling person opens the door and the band, still playing, half-dances, half-walks into the home. Most homes in Anamat are built on platforms, and in those homes with wood floors, the floorboards themselves start to bounce and dance, with the dancing of the *paranderos* crowding into the living room. After the first song, everyone takes a drink and wishes the people of the house "Merry Christmas." After the first round of drinks, the band begins another song. This cycle might continue several times before the band plays its final song as its members dance out the front door. Often, a visit to a house results in new members for the chorus. At around eleven at night, the size of the band peaks. The largest band I witnessed counted well over seventy people who filled the living room, front gallery, and some of the steps.

According to *paranderos*, a "real paran'" must continue until the sun rises. Soon after midnight, the size of the band decreases, but the core *paranderos*, the "true *paranderos*," continue. Another "must" is that every *parandero* must drink something at every house. After midnight, hosts supplement their offerings of alcoholic drinks with coffee. Aside from some cake, food is rare, so as the *parang* evolves, the stomachs become empty, but more alcohol and caffeine are absorbed into the bloodstream. As the night wears on, the alcohol and caffeine combine with the hypnotic ssshhhk-ssshhhk-SSSHHHK of the shak shaks, the never-ending clink-clink-clink of someone beating a bottle with a spoon, and the droning of the stringed instruments, to generate a state of trance. With each new night, after *paranging* the night before, the trance state is achieved more quickly. The division of day and night becomes irrelevant; one might begin to *parang* at four in the morning or at nine at night. One might still be *paranging* in the middle of the afternoon. Those who receive, rather than give, *parangs* learn that no time is immune to *parang*—neither sun nor stars determine when a band might arrive. Conventional notions of time are suspended; time is punctuated only by the sequence of songs and the movement from house to house. Even these two experiences become blurred as *parang* season wears on: after several nights of *paranging*, for *paranderos*, Christmas becomes reduced to a rhythm of movement and song, in which the beginning has been forgotten and the end comes only when one's body gives out. My own threshold was reached on the second night, but experienced *paranderos* seem to fall into this trance earlier than I did, and they seem to be able to maintain the pace of nightly *parangs* for more nights than I was able.

In effect, *parang* contains a distinct means of marking time (movement from house to house), as well as an attempt to overcome natural cycles of day and night. The ability to transcend time in this way is a source of pride for *paranderos*—a triumph of the human will and music over nature.

### Carnival and Time

For many Anamatians, Christmas is a prelude to Carnival. During the first week of January, the first large-scale fetes are held. Such fetes consist minimally of a disc jockey with a large sound system. The larger fetes, usually sponsored by large corporations such as Amoco and Nestlé, include several bands and calypsonians, and they draw revelers in the thousands. Most Anamatians who participate in Carnival attend several of the corporate-sponsored fetes leading up to Carnival, and possibly the national steelband competition known as Panorama. They then "meet" Carnival early J'ouvert Monday morning in Port of Spain. J'ouvert is the official beginning of Carnival. Anamatians associate it with "old mas." This form of mas involves inexpensive costumes, such as the men who dress as women (*Jamettes*) and individual masqueraders who develop costumes around political themes, such as the Persian Gulf conflict over Kuwait during the 1991 Carnival. The most popular J'ouvert mas among Anamatians is mud mas, in which people cover themselves with mud and dance through the streets. These types of mas are a contrast to the expensive, elaborate costumes of the masquerade bands that dominate most of Carnival.

Just before the sun peeks over the horizon, most revelers from Anamat leave J'ouvert in Port of Spain and head for Arima to greet the sunrise with steelbands. While Port of Spain has many steelbands, on J'ouvert they are always easier to find on the streets of Arima. Finally, around midmorning, the "posses," as the groups are called, return to Viego Grande, the market town nearest Anamat, for the remainder of Carnival.

Trinidad's Carnival defies easy description. Trinidadians participate in Carnival in many ways: watching the event on television; attending fetes every night; "playing mas," that is, playing a masquerade by marching and dancing in costume; and liming alongside parade routes. Some participate from J'ouvert early Monday morning until "last lap" at midnight on Tuesday, whereas others may participate for only part of one day or not at all. Trinidad's Carnival, like many festivals, plays with the structure of everyday life. Trinidad's Carnival is colorful, loudly musical, and sensual. During Carnival, Trinidadians take over the streets. Some of those occupying the streets are members of mas bands dressed in stunningly colorful costumes. Others prefer mud mas. Still others prefer the role of active spectator—dancing and jumping to the music during the day and night. For some, Carnival is recreational. It is an occasion to relax one's inhibitions and "get on bad"—in other words, to drink, dance, and, if one desires, seek sexual liaisons. During the Carnival of 1991, the icon of getting on bad was a photograph on the front page of a daily newspaper of a female teacher waving a pair of panties during a performance of the song "Get Something and Wave." For others, Carnival is a time of satire where the mighty and the powerful are ridiculed through song, costume, and subversive behavior.

The Carnival fetes Anamatians attend lead up to the climax of the festival, which for Anamatians is J'ouvert. These fetes vary from small affairs, sponsored by local bars and clubs, to large, corporate-sponsored events with thousands of revelers. Calypso and soca are provided by disc jockeys and live bands.[2] Many popular soca artists make guest appearances, singing their songs to crowds dancing and drinking in front of the stage. Fetes are usually bounded in space by an enclosure, into which there are limited entrances at which one presents a ticket. The fetes often begin as early as eight o'clock, but the scheduled beginning and the rhythm of the fete are two different matters entirely. The crowd is small and calm at eight. Gradually, the crowd grows until it packs the fete enclosure, sometimes with thousands of people at the larger fetes. Since the law requires police to be present when liquor is served, and since the police go off duty at 4 A.M., most fetes end at this time, but not out of choice. Almost every year, there is at least one song about a never-ending party, inspiring fantasies of unending revelry. At fetes, one's arrival and departure are indeterminate and unscheduled.

Carnival synthesizes linear and cyclical metaphors of time. The same progression of events and fetes occurs every year. All the fetes contain similar structures of activity giving them a cyclical quality. At the same time, the fetes and competitions build up to Carnival's climax, which for Anamatians is J'ouvert. Moreover, while the cycle of events is repeated every year, the lyrical content of the calypsos and the costumes and themes of the masqueraders change, often capturing recent newsworthy events (see Anthony 1989).

Anamatians look forward to particular fetes every year. These include the corporate-sponsored fetes, such as the Nestlé fete, the Carib fete, and the Amoco fete. Since these fetes are held at the same time relative to Carnival every year, and since their order does not change from year to year, it is easy to recognize which fetes will be held each weekend during the Carnival season, even without advertising. Plans can thus be made to attend a fete far ahead of time. Even if such a trip is planned, there is little organization in the time to leave. Those planning to attend might begin to gather at approximately eight o'clock. When the driver is ready, everyone squeezes into the car, and the posse leaves Anamat. Since the roads from Anamat lead to Viego Grande, the driver often stops there to buy gas, and the posse buys the first round of drinks, usually beer. After liming a short time at the gas station, the posse again mobilizes and heads "down the road" to the location of the fete. Most fetes are held in the environs of Port of Spain. Along the road to Port of Spain are several snackettes, at which the posse will stop and purchase additional rounds of drinks. By the time the posse arrives at the fete enclosure, everyone has consumed two to three beers. The reciprocal buying of beers serves to cement the social ties of those in the posse, and the willingness to drink together is something that, in Trinidad, is associated with asserting equality and friendship (see Angrosino 1974:32–33).

The behavior of the posse members shows that it is not organized according to any temporal coordination. There is no agreed-upon time to meet at the fete. They use no watches or clocks to coordinate their activities. Instead, each car arrives whenever the driver wants, and the people in the car utilize strategies dependent upon space, rather than time, to find friends. Indeed, nobody's personal temporal model is relevant. The car does not leave Anamat until it is full. It arrives at the fete after a journey of several stops of unequal length. The only temporal constraint at a fete consists of the ending of the fete at 4 A.M., because that is when the police, who must be present when alcoholic beverages are served, go off duty.

Since time, particularly a time formally defined by institutions, is important to the coordination and even the subjugation of persons, festival time poses a threat. This threat does not go unchallenged. Indeed, while Carnival exists in contrast to the formal definitions of time associated with employers and the government, embedded within Carnival is the temporal response of government and employers to the festival. Carnival threatens their power over people through control of time. Those who have vested interests in maintaining the status quo also have interests in controlling Carnival. The National Carnival Commission (NCC) has been an entity through which the state has attempted to control and appropriate Carnival since the late 1950s (Stewart 1986; Stuempfle 1995). The NCC schedules Carnival events and attempts to choreograph the number, pace, and route of performers by making all bands, both mas bands and steelbands, follow a parade route that includes a stage area. The amount of time a group is in the stage area is carefully defined by the NCC and is openly defied every year by mas bands. While these strategies move to control Carnival temporally, they do not work very well. The frame of Carnival makes all temporal models irrelevant and undermines the effectiveness of the NCC's efforts: bands stay too long on the stage area or delay entering the stage area. This temporal battle is waged every year, and the NCC loses every year. For this reason, the NCC's attempts to exert temporal control over Carnival cannot be viewed as effectively exerted political power but must be seen, instead, as an effort of those in power to resist and control the disorder of Carnival.

In Trinidad's Carnival, no time matters; festival time features an irreverence toward normal times. Whatever time someone may normally employ does not matter during Carnival, because no time is respected. During Carnival of 1991, the Carnival following the attempted coup d'état and its associated curfew, the most popular soca was "Get Something and Wave" by Superblue. This song featured the chant of "no curfew, no curfew," as well as the line addressed to law enforcement, "I ain't kill, I ain't break no law, what you timin' me for." Thus, rather than manifesting a definition of time, festival situations manifest an institutionalization of a call to ignore or challenge all definitions of time—no time can be taken seriously; no time can be enforced. It is inappropriate to hold people to any sort of schedule or rhythm during Carnival. In this manner, farmers, taxi drivers, and factory workers all come together. They must tolerate one another's times and heed the decree of the calypsonian Lord Kitchener, who sang in the 1970s that "Any time is Trinidad time." Moreover, for Anamatians, Carnival is like *parang* because they both defy circadian rhythms by negating the distinction between day and night, encouraging activity when one would normally be sleeping, and promoting irregular eating habits. It undermines one's biological clock, rather than affirms it.

Time during Carnival is different from *parang* in some ways, however. Whereas *parang* involves transcending day and night as an important source of pride, Carnival undermines the legitimacy of all times—biological and social. It is not enough to say that the time during festival periods is opposed to everyday temporal models, however. Since everyday life consists of multiple times, festival time is in opposition to multiple times. For festival time to subvert everyday times, it must deny them or ignore them. There is no pride in playing mas until daybreak—the pride emerges from the fantasy of never stopping—of a continuous negation of everyday times. Such is indicated by the fascination with such themes in popular socas. Tambu's "No, No, We Ain't Going Home," the most popular soca of 1990, deals with a party that never ends. Superblue's 1993 hit, "Bacchanal Time," suggests that every day of the week is "Bacchanal Time" and should be devoted to Carnival and partying.

Thus, Carnival and *parang* avoid temporal conflicts by negating temporal expectations—a very effective means of doing so. There are no problems of enforcement and no need for sanctions within Carnival, but, at the same time, Carnival subverts the temporal models that allow social relationships to be coordinated with one another. For the Trinidadian social system to function, Carnival time cannot be the dominant time all the time, yet given the temporal confusion generated by Trinidad's diversity, it is not surprising that there exists a context in which ignoring temporal difference obtains social significance.

## Conclusion

This chapter has explored various ways of coordinating social relationships in time. The descriptions from Trinidad do not provide an exhaustive classification of such ways but instead provide examples of the different means in which this occurs. Temporal coordination is achieved in many ways and is not limited to the application of power. In formal institutions, power is utilized. In contrast, during Carnival no individual or institution is given the power to define time. Indeed, on festive occasions, participants defy all systematic definitions of time. The application of a universal time can involve consensus or the exercise of power, but a central authority is not necessary for it to function.

No means of defining and enforcing time is completely effective. The first, institutionally defined time, involves the expression of power through coercion and the recognition of authority through legitimacy. It is also the means of defining time that is most likely to be contested—such as the foremen's tricks with their watches. The second means of defining time and avoiding temporal conflict, the means probably most familiar to readers, is the use of a universally recognized time. In many cases, this involves the use of clocks and calendars to set a temporal standard. This standard can be related to other ideas of time. Like institutionally defined time, the use of universally defined time is met with opposition. In addition, there are cases where it is assumed that everyone is familiar with the standard, but everyone is not: not everyone can read a watch or judge time by the position of the sun. Third, festive time is effective because of the irreverence toward the definition of time found in festive contexts. Festivals generate an "anti-time," in that they call revelers to disregard any attempts to define time. Yet, this way of avoiding temporal conflict does not sufficiently coordinate daily life in a complex society. This is limited by the complexity of Trinidadian society. Consequently, while there are many means to achieve temporal coordination, none is effective at all times and in all contexts. Much of daily life occurs in situations in which there are no clear means of defining time and avoiding temporal confusion.

# Cohesion in Chaos

**5**

Many everyday social relationships unfold outside the boundaries of institutional definitions of time. Where no formal definition of a time exists, the burden of managing the problem of temporal differences falls on the people involved. In Anamat, differences in the rhythms of production and social interactions form part of the social fabric of everyday life. When differences occur, individuals must decide between overlooking the differences or emphasizing them. The former encourages social cohesion, and the latter encourages social conflict. In situations of conflict, temporal differences become part of larger conflicts, such as those between genders, generations, classes, or ethnic groups. In fact, temporal conflict is an important means of expressing these other conflicts.

In Trinidad, the words or phrases used in these situations of temporal differences are examples of "chat" or "slang." According to Trinidadians, to "give chat" or to "use a slang" is to obscure deliberately the truth. As one Creole explained, "We always have head work, because we don't speak straight, you know? We have many different words used instead of straight language. Instead of saying 'yes' or 'no,' we tell you a lot of different things. It means 'yes' and 'no,' but we have to work it out to know what we mean."

One use of slang is as tokens. Tokens are used to suggest agreement where there is actually conflict (Swartz 1984, 1991). This allows persons to minimize overt conflict without compromise or capitulation. For instance,

Michael was a young adult who despised agricultural work and enjoyed liming. He epitomized the behaviors and attitudes that old heads mentioned in their complaints about youths. These complaints included references to a lack of respect for time. Yet, in interviews, Michael took the position of temporal discipline usually manifested by old heads. He said,

> We must be more serious instead of [acting] like we don't care. Like, for instance, about the time and thing, we must be more serious for that. Like, if you keeping something, and you say that the time you keeping it, try and make the system start on that time. For instance, look at the sports [that was held in the village last weekend]. People come out real early for that sports, and a real big turnout real early and things didn't start until around three o'clock [two hours after the announced starting time of one o'clock].

Later, Michael commented,

> Everything never start on time here in Trinidad, never start on time. Everything jus' wrong. I don't like how everybody carry a different time. They don't care. [In] Trinidad, like everybody ain't care in truth. They really ain't care in Trinidad. I, for instance, I does catch myself saying the same thing, "I don't care." Well, that isn't to say I don't care, but I jus' say that because that is the slang.

He ended with a judgment about Trinidadian youths that echoed what old heads said: "The majority of youths now born and grown with that, so them the same thing that I be saying." So Michael condemned the same behaviors that the old heads condemned, and he admitted that he was guilty of such behavior. In effect, this was a token agreement with the old heads about time. Tokens have potential costs to them: to assert that one thing is true, and then to behave as if it was not, has negative implications for the public evaluation of one's sincerity and morality. Michael feels those costs. While he is liked in the village, landowners with work opportunities do not employ him. So much so, that he felt that he needed to leave Trinidad in order to make a new start.

Tokens are not the only means by which conflicts are mitigated. Other phrases are designed to obscure conflict rather than to pose a superficial agreement such as that accomplished by tokens. This strategy for dealing with different times involves what I call "glosses." These obscure differences thereby prevent confrontation with contradictions in one's behavior. The use of the phrase "any time is Trinidad time" is an example. Since any time can be Trinidad time, then all times are Trinidad time, according to the logic and use of the idiom.

Both forms of subterfuge, tokens and glosses, fall within the Trinidadian category of "slang" or "chat." Yet, these do not exhaust the resources available for Trinidadians to overcome temporal conflicts. There is one more strategy that overcomes temporal conflict. When there is a conflict between two individuals because of different rhythms or conceptions of time, one individual may change the time he or she follows to create a semblance of consistency. The frequent shift that farmers who also work for the government make between clock time and the daily cycles of agricultural work is an example of this.

## Dimensions of Dispute

Many temporal disputes concern punctuality, but, to my knowledge, all previous discussions of punctuality have focused primarily on clock time without considering how punctuality could be defined in other temporal systems. Punctuality is quite variable and need not be determined by reference to a clock. In fact, punctuality is more closely tied to social rhythms than to clocks. When people associate punctuality with clocks, it is because clocks are used to structure social activities. In contrast, E. E. Evans-Pritchard makes a point to say that Nuer time is not defined by a clock. He points out that time is reckoned by the diurnal cycles of tending to cattle (1940: 101–2). In effect, the Nuer still expect certain tasks to be done within particular temporal parameters, but those parameters are defined not by the clock, but by the requirements of tending to cattle.

Punctuality, as I am using the term, involves arriving at the correct, appropriate time, but such correctness is culturally determined. In some cases, people make culturally constructed links between natural and social rhythms, and these links form the basis of determining punctuality. Cutlassing, the cutting of bush in cocoa walks, is one Trinidadian example of punctuality being determined by means other than clocks. Cutlassing is best done in the evening or morning, when the bush is softer and more easily cut. A worker who shows up when the sun is high and the air is warm will not be able to cutlass as quickly as a worker who shows up when the sun is low and the air is cool. Employers expect workers who are cutlassing to start work early in the morning because of this physical constraint of cutting grass. So, with regard to cutlassing, punctuality is not determined by a clock but by the position of the sun and the heat of the day. Lateness, then, is defined by the position of the sun and the temperature, not by the clock. To show up "in the hot sun" to cutlass is interpreted as lacking the motivation to do the work. At a rum shop one evening, a landowner was complaining about a man he had hired to cutlass the bush in his cocoa: "The man show up at noon, take two swings with his swiper (cutlass), and say, 'This bush too hard,' and asked for more money!" The landowner then explained how he told this worker not to bother coming back to work.

In a horologically obsessed society, punctuality seems to require appearing at the correct moment as defined by the clock. Even this is not completely true; in some contexts, such as at parties, it is appropriate to be "fashionably late," and being "on time" in this kind of situation requires being late according to a clock. Indeed, arriving on time in some situations is socially awkward. There are cases when being "on time" in Trinidad involves tardiness. Fetes are just one occasion when to arrive at the predetermined clock time is to arrive at an inappropriate time. When the social rhythms of fetes are taken into account, arriving two hours late is more appropriate than arriving at the scheduled starting time of the fete. Trinidadians prefer to arrive after the event is under way. As one informant explained:

> It doesn't matter . . . When you make an appointment [to go to a party], we ignore it, because, you know, if a party is advertised from eight 'til four, the party ain't getting under way until about 10 P.M., so we don't make a specific time so that when we going to a party that we make it on time, ahead of time, but when the party ripe . . . We want to get into the party when the party ripe. We don't want to go and waste no time, see an empty hall.

Basically, Trinidadians prefer not to sit and wait for a fete to become interesting—especially when one can be doing something else. Most fetes are at night, and their scheduled starting time ranges from 9 P.M. to 10 P.M. For Trinidadians, the daily rhythms of eating meals and watching television delay preparation until around 8 P.M.—after the seven o'clock news. Preparation involves bathing as well as dressing. Consequently, most Trinidadians do not leave for the fete until about two hours after they say they are going to leave. Since this approximately two-hour delay is so widespread, it functions to synchronize activities as effectively as, if not more effectively than, a universal time reference, such as a clock. The personal rhythms involved, such as individual cycles of eating, sleeping, working, and other daily tasks, are not dictated by the clock, but because of widely shared values, the personal rhythms are very similar. Thus, when the emergent social rhythms are taken into account, to be punctual actually involves arriving about two hours after the formal starting time.

Thus, punctuality is related to the sort of activity in which one is engaged. Trinidadians are not punctual relative to clock time in their attendance of fetes, nor should they be, because clock time is not the relevant time at fetes. For some tasks, Anamatians determine punctuality by natural factors, such as the fact that it is easier to cut bush in the morning than in the heat of the day. If they are hired as wage laborers, they are punctual by clock time in arriving for work, however. For instance, one man always left the village to go to a fete long after the time he announced he would leave, but he also took pride in the fact that the only time he was ever late to work when he worked at the Amalgamated automobile assembly plant was when his car broke down on the way. One might say that he exhibits two very different attitudes toward time, but another way of thinking about his behavior is that he exhibits at least two different times with two different definitions of punctuality.

While it is impossible to lack punctuality during Carnival, it is possible to be late to fetes the rest of the year, particularly those in Anamat. Again, tardiness cannot be defined by the clock, but by the social rhythms inherent to the context. In the case of fetes, Anamatians associate a culturally constructed cycle of "hot" and "cold" with time. They discuss fetes in terms of their temperature. One does not want to be at a fete when it is too cool or too hot. The cooler a fete is, the more boring it is. Coolness is marked by small numbers of people, little dancing, poor music, and no threat of violence. Hot fetes are rife with sex, drunkenness, and violence. The Trinidadian cultural model of fetes contains a script in which fetes tend to begin as cool and then get hot. The appropriate time to arrive is when it is "getting hot." One should then stay while it is hot but leave before it gets too hot—too hot is indicated by too many fights and bottle throwing.

The relativity of punctuality, an implication of multiple times, sets the stage for temporal conflict. It is possible for two individuals to be in agreement about the importance of punctuality and for them to arrive at an agreed meeting at very different moments. While all times are potentially referable to a universal time—for example clock time or the position of the sun—since universal times are not directly integrated into many contexts, their use can result in conflict.

A second temporal conflict arises over meanings attached to a time period. Activities and time periods can be intertwined. A classic example is the relationship between labor and time when workers are paid an hourly wage. In this situation, workers are technically paid for their labor power, but their labor is valued in terms of hours and minutes. Consequently, the hours during which they work are equated through wages with the tasks involved in work. Workers are aware of this and adjust the amount of work they do to the time they are expected to work. Kevin Yelvington provides an example of this in his description of a Trinidadian factory. He describes how one set of workers is expected to produce 200 items a day by the owner of the factory. The workers feel that this is too much, and they set their quota for 150, and they produce 150 daily, keeping "close count on the number produced" (1995:202). By 4 P.M., a half hour before the end of the workday, they have usually fulfilled their self-defined quota. Yelvington argues that the reason they set their quota at 150, instead of at the management's 200, is that they know that if they produced 200, the management would increase the quota (1995:202). Regardless of how much they produce, their wages remain the same, although, theoretically, producing too little would jeopardize their wages, and even their jobs. In any case, because the wages remain stable despite fluctuations in how much they produce, they set their production quota to ensure that the owner will not expect too much of them. In determining wages for piecework, employers still often begin from the calculation of the wages for a day's labor and then determine the wage per piece so that adequate production will result in these wages.

Such connections between time and activities are not limited to work situations. Many religious traditions assert that activities and times are intertwined. Hinduism defines particular days and times as auspicious, thereby defining when it is advisable or potentially dangerous to begin business transactions, to marry, or to construct a home. These times are determined by pandits and recorded in almanacs that Trinidadian Hindus purchase. Within temporally defined limits, activities and actions are deemed appropriate or inappropriate, and these evaluations can differ and become the source of conflict.

### "Any Time Is Trinidad Time" and "Time Is Time"

"Any time is Trinidad time" has become a token expression (see Swartz 1984, 1991) that conceals temporal disagreements by suggesting agreement through "chat." Trinidadians use the phrase "any time is Trinidad time" to explain a lack of regard for universal times in general and clock time in particular:

> Like you going down the road, and I tell you like you have an appointment with me and, you tell me, "Look, it nine o'clock in the morning." You might come nine o'clock and you mightn't meet me. You meet me [later], I say, "Well, to me, you tell me nine o'clock this morning, man, 'any time is Trinidad time'," you know: "any time you reach you reach."

Anamatians contrast "any time is Trinidad time" with the assertion that "time is time." They share an idea that among the Americans and British "time is time," and, consequently, the phrases imply a contrast between ideas of Trinidadian and American/British temporal attitudes. This shows the power and extent of American and British influence concerning cultural ideas of time. The way Anamatians draw the contrast between "time is time" and "any time is Trinidad time" suggests that they view "time is time" as morally and socially acceptable, and "any time is Trinidad time" as morally and socially questionable. In discussing Trinidadians' use of "any time is Trinidad time," one man told me, "They don't care. They don't think maybe of what they are really saying." He then went on to suggest that these individuals are shirking their responsibilities:

> So when they use "any time is Trinidad time," I believe some of them does mean anytime they feel to go. They aren't thinking, "At

this time I have to go and pray" [or,] "At this time I have to go and work or something."

In effect, the use of this idiom suggests self-centeredness and a lack of caring in contrast to the idea that "time is time." This latter idiom indicates responsibility.

Anamatians view those in authority who use the phrase as abusing power. One of the complaints that Trinidadians commonly make about government offices is that government personnel do not act like "time is time," but like "any time is Trinidad time." In the case of "time is time," one arrives at the appointed clock time or earlier. To do so shows "caring" and "consideration" for the needs of others.

As I discussed in chapter 1, the association of clock time with the United States and Great Britain developed before independence. Under the colonial administration, many supervisors in the government and on plantations were English, and they held their workers to a model of time as defined by the clock. Furthermore, during World War II, the U.S. Army employed several thousand Trinidadians and also enforced clock time. The links between clock time and work are still powerful, but many old heads report that there is more lenience now than there was under British rule or at the American military bases. In their view, independence and the power of unions have resulted in a laxness with regard to time and an inability to punish those who are not punctual. One commented,

> Now, the discipline long time [ago], [it was such that] the chief overseer on the road could send you back home [if you were late to work] and you would have to lose a day work. If he do such a thing now, he will be answerable to the union. In his own discretion, he is right. He send you back home. Next time you make sure you come on time.

These old heads further blame youths for adopting a more relaxed attitude toward time. One said, "That has become a way of life in some people, and the younger ones seem to be exploiting it because they see too much of the younger people going to work at any time." As a result of their descriptions of life under the British, both youths and old heads come to identify "time is time" with British colonial rule. To a great extent, the historical accuracy of such an identification is no longer important; it is the association between the two in contemporary discussions that makes the connection important. The assertion that "time is time" in England and the United States is further supported by Trinidadians returning from working in either the United States or England. These assertions do little to change conceptions of time employed in Anamat and serve primarily as a basis for contrasting Trinidad with "foreign," a term used to encompass England, Canada, and the United States.

My first encounter with the phrase "any time is Trinidad time" happened soon after I arrived in Anamat. I was invited to a Bible study. I was told the time the study was to start, and I arrived on time to find that many of the participants were still absent, including the study leader. Much later, the leader arrived. He apologized to me and began to explain how "any time is Trinidad time." I asked what he meant, and he explained that, while Americans live by the clock, Trinidadians do not. Instead, he said, Trinidadians come and go as they want: they come any time, and they go any time. On another occasion, I was told that "any time is Trinidad time" applies to "any purpose where we [for example] make a plan to start the game, let we say a cricket game, at seven o'clock [in the] morning, but we wouldn't actually start seven o'clock. We start eight o'clock and say 'any time is Trinidad time.' We really say seven, but really ain't mean seven." This informant added later, "Still, for all, if you reach on time, still you get the chat—they tell you 'any time is Trinidad time.'"

As described earlier, the purpose of "chat" or "slang" is to "not come straight," meaning to obscure the truth. In using "any time is Trinidad time," individuals attempt to gloss over the fact that they have arrived at a time potentially deemed inappropriate by another, although not necessarily by themselves. Those who use this chat recognize temporal conflict and respond with the chat that provides a means of mitigating the conflict without apologizing and without changing one's behavior. Embedded within the phrase "any time is Trinidad time" is the implication that Trinidadian nature defies adopting a strict sense of clock time and punctuality defined by clocks. That was what the Bible study leader attempted to convey to me. He knew that I had just arrived in Trinidad, and he tried to explain what he thought to be an aspect of Trinidadian character.

This phrase gains persuasiveness, because it evokes ideas that are part of Trinidadians' view of themselves. Trinidadians do not contest the veracity of the ideas that lurk behind the phrase, but they do dispute the moral value of these ideas. Some see the view of time associated with "any time is Trinidad time" as positive, because it reveals, using their word, the "freedom" Trinidadians have—"you see, 'any time is Trinidad time' is true, eh, with freedom . . . It mean, like, freedom—anytime you go anywhere." For Trinidadians, the idea of freedom involves a lack of constraints on behavior and expression (Miller 1991), particularly behavior that defies their conception of European, particularly English, morality. Trinidadians assert that they are the freest people on earth. Repeatedly, Trinidadians told me that they had more freedom than "even Americans." They base this claim on what they hear from emigrated relatives and the various media from the United States disseminated in Trinidad, such as popular music, television, and the Cable News Network (CNN). Moreover, Trinidadians are aware of the claim by American leaders that the United States leads the world in freedom and democracy. Trinidadians respond to this with a sort of one-upmanship, by claiming to have greater freedom than Americans do. Trinidadians see the lives of Americans as ruled by the clock, whereas Trinidadians assert that, for purposes of this contrast, they are free from the constraints of the clock.

Since temporal inflexibility is associated with European or American times, the assertion that any time could be a Trinidadian time is an assertion of difference from temporal ideas thought to be European or North American. Thus, it affirms a unity of Trinidadians in relationship to others, particularly those who had the most power over Trinidadians under the colonial administration. Interestingly, however, Trinidadians use this affirmation of Trinidadian solidarity also to obscure differences with others. A visitor from the United States or England who gets told that "any time is Trinidad time" most likely will not realize that this seemingly innocent, self-deprecating commentary on Trinidadian character is really a veiled assertion of Trinidadian solidarity in response to American and English cultural imposition.

Other Trinidadians disparage the implications of "any time is Trinidad time." They claim this phrase as evidence of Trinidadian moral failings that underlie the island's economic underdevelopment. They argue that "time is time," and that it is impossible for "any time to be Trinidad time," suggesting that "any time is Trinidad time" is simply an excuse for laziness and mediocrity. They assert that "any time is Trinidad time" obscures the reality of time and takes too much liberty with time: "Time is time, not just any time." Still, this associates "any time is Trinidad time" and its latent content with freedom. They say that Trinidadians claim too much free-dom—even freedom from nature, as shown by the claim of freedom from time. Furthermore, they see this as a moral failing that has political and economic implications, leading not only to disrespect in social relations and low efficiency in the workplace but also to widespread corruption. Those critical of the ideological underpinnings of "any time is Trinidad time" view freedom as having a degenerative effect on Trinidad's society. They feel that, long ago, when people were not as free (but after slavery and indentureship), the island was much better off. Therefore, there is no disagreement that Trinidadians are free, but there is disagreement over whether this is moral or good for the country.

Thus, when someone says, "any time is Trinidad time," there is little, if any, disagreement as to whether this is the case or not, although the people hearing the remark may strongly feel that it should not be case and take punitive measures based upon this view. Indeed, if they have the power to take such action, they are expected to do so. For instance, employers are expected to fire or to reduce the pay of workers who arrive late. In such cases, on the basis of this appeal to Trinidadian character, the temporal transgressors ask for leniency. The "slang" states that the transgressors were not late by their own volition, but that their character as Trinidadians caused them to be late. As a token, therefore, "any time is Trinidad time" affirms an agreement about Trinidadian character, however that may be evaluated: "We, as Trinidadian, really can't vex because we have too much chat."

There is another dimension to this appeal, however: slangs or chats usually imply levity. The use of slang in a potentially tense situation is designed to relieve the tension, thereby serving to correct whatever social wrongs may have been committed. In addition, to appeal to an image of Trinidadian character, the use of the slang or chat "any time is Trinidad time" is the deployment of humor to diffuse disagreement. As one young Creole man said, "They jus' use it as a joke. I don't know if they serious about it or they jus' does use it, maybe. I know, for instance, sometimes you might go to play a cricket match, like a friendly game starting at one o'clock. You reach there half past one, two o'clock, and it ain't start yet. When it start, any time is Trinidad time."

If the other person responds with humor, the chat was successful. The success of chat depends on those involved in the interaction. This seems to be a West Indian pattern: Roger Abrahams (1983) discusses the importance of the clever use of language as a means of establishing and protecting one's reputation, and Brackette Williams (1987) adds that verbal virtuosity is an important component in the resolution of disputes in Guyana. Brian, a man who works on his father's land and occasionally on temporary jobs for the government, provided an example of the effectiveness of chat in some government work: "It have times when I was working, sometimes it reach to ten o'clock [three hours after the starting time]. [Someone asks me], 'You ain't going to work, boy?' I say, 'Yeah man, I going jus' now. You ain't know any time is Trinidad time?'" Brian then said that when this happens, he might get fired, but sometimes he is simply told, "Like you is your own boss." He later commented, "Most of the time, you reaching after seven [the starting time for work], but you reaching before the boss, so most of the time one does say 'any time is Trinidad time.' Sometime, the boss ain't reaching later than you, but the boss really ain't setting an example towards people that working [i.e., coming to work] late." If a chat is successful, it suggests a degree of egalitarianism between the two people involved, or that the individual who deployed the chat has power over the person who accepted the chat. One informant told me, "Trinidad people get so accustomed to it ["any time is Trinidad time"] that you will tell them, 'Trinidad time is any time' and they just laugh it off and they call it good." Bosses who show up to work very late use the slang "any time is Trinidad time" to justify their behavior. In these cases, workers have to respond in good humor, because the bosses have power over them. Likewise, in such situations, if a worker arrives late and uses the slang, and the boss accepts it, it symbolizes equality between worker and boss. Generally, such a scenario only happens if the workers and bosses lime together outside the work situation.

Finally, this "chat" obscures whether the lack of punctuality arises from temporal differences or inexcusable tardiness. In doing so, it is clearly a gamble in which forgiveness might be gained or resentment provoked. Commonly, the receiver of the chat will act tolerant but harbor resentment.

In effect, three different sorts of strategies to diffuse tension are embedded in the phrase: an appeal to Trinidadian character, an appeal to humor, and an appeal to recognizing and tolerating difference. Its use is a deliberate attempt to avoid retributive action and to correct social damage. As an appeal to freedom, humor, and equality, it is an appeal to three fundamental components of the image Trinidadians have of themselves.

The opposite of "any time is Trinidad time" is "time is time." This phrase also serves as a token. In effect, Trinidadians use the two phrases with opposed meanings as tokens, and their use depends on whether they want to take a stance siding with the colonial definition of temporal morality and declare "time is time," or whether they want to side with an anti-colonial morality and Trinidadian solidarity and assert that "any time is Trinidad time." It is clear that older Anamatians use the phrase "time is time" much more often than young Anamatians. Indeed, the phrases "time is time" and "any time is Trinidad time" are embedded in a conflict between age cohorts. The old heads in Anamat tend to use "time is time" to condemn young Anamatians. In interview situations, the young Anamatians affirm that it is true that "time is time," but that they do not act as if it were. When, usually over beer in a local rum shop, members of both groups come together and lament the condition of the road, or the economic recession, or the country as a whole, they often take the position of "time is time" and decry how Trinidadians are ruining their country. A few old heads even hearken back to colonial rule and argue that everything was better then, because there was greater respect, and a few openly wish for the United States to make Trinidad a colony. This fits with the results of the island-wide survey issued by Selwyn Ryan, Eddie Greene, and Jack Harewood. One of their prompts was "Trinidad and Tobago would have been better off had it not become independent": 53 percent of the respondents agreed with this statement (Ryan, Greene, and Harewood 1979:21). Depending on the speaker, the young adults in the audience politely agree with an old head's opinion, often with flattery about how the old head has seen so much in his life and knows so much. Such acquiescence does not suggest that the young adults then adjust their behavior to conform to what the old head thinks is right, or that they wish for a return to colonial status; quite the contrary, they do not wish for either one. But, because the context is one of liming and is supposed to be "cool," they will not challenge the old head or provoke conflict. The token "time is time" serves its purpose of preventing temporal conflict.

These two temporal tokens, "time is time" and "any time is Trinidad time," serve their purpose of preventing temporal conflict by appealing to shared sentiments. The former appeals to morality, and the latter appeals to a feeling of shared Trinidadian character. Of the two, the latter is the most widespread and seems to be the more dominant of the two through its appeal to anti-colonial ideas, freedom, and humor. Even though it is the more pervasive of the two, this does not mean that it is deemed appropriate in all situations.

### Glosses and Obscuring Time

Another means of avoiding temporal conflict is to obscure time distinctions. Again, this is accomplished through a form of chat. In addition to "any time is Trinidad time," which can be used as a gloss, the two most common "chats" used as glosses are "jus' now" and "long time."

Trinidadians use "jus' now" when they do not want to be bound to the clock to finish a task. "Long time" sometimes serves to suggest quickness and efficiency in completing a task when neither one has been necessarily the case. The use of these terms pervades Trinidadian society, and their flexibility results in their being adaptable to any temporal system.

"Jus' now" indicates an imperative "wait." Normally, the waiting lasts only for a few minutes, but sometimes it indicates much longer periods of time. This use of "jus' now" tells a person to wait until a task is quickly finished. This is evident from the examples Trinidadians give. As a Creole told me:

> You can be writing a letter. You studying something to put in the letter, and you thinkin' how to write this—and he calling you. You say, "Wait a while now, man—I comin' jus' now." You find you finish up. When you done, you gone.

Or, as an Indian explained: "Perhaps I might be under the vehicle, you call me, and I want to complete something under there. When that come through and I complete that, the 'jus' now' might reach." Thus, "jus' now" sends a message of "let me finish what I am doing first, and then I will be with you." As a result, "jus' now" is often profoundly embedded in a task orientation of work, and since many types of work have inherent rhythms, the meaning of "jus' now" varies according to the work involved. Since the meaning of "jus' now" is purposefully vague, it effectively serves as a means of negotiating between two or more times. It signifies a priority of tasks in the order in which they are demanded—the second task is undertaken only when the first task is completed. The use of "jus' now" is not limited to mitigating conflict between a universal time and other times; it is applied when any two times are not synchronized. For instance, the small shopkeepers in Anamat make frequent use of the phrase because they usually work in their shop at the same time they complete household tasks. When a customer arrives, the shopkeeper will answer, "I comin' jus' now." Whenever "jus' now" is used, it blurs the differences between times.

The use of "jus' now" is not limited to situations of conflict. Trinidadians also use "jus' now" to shrink time rhetorically. For instance, on January 8, the day after Christmas officially ends, someone might say, "jus' now is Christmas again." Trinidadians use this statement to emphasize two things: how quickly time passes and the importance of the event. To say "jus' now is Christmas again" makes Christmas seem much closer than saying that "next Christmas is a year away." Also embedded in this usage is an emphasis on preparedness. To say "jus' now is Christmas again" communicates that next Christmas is coming soon, and that one should get ready. Usually, it is *paranderos* who say "jus' now is Christmas again." This indicates that those who participate in the event in a significant way make use of this chat.

How is this second usage related to the first? In both cases, the phrase obscures duration—particularly time spent waiting. To say "jus' now is Christmas again" serves the same basic function as telling someone "I comin' jus' now." In both cases, the audience is being told to become, or to remain, prepared and to wait. In both cases, duration is left ambiguous. Trinidadians use the former usage, "I comin' jus' now, " to avoid attaching a specific duration to the time one must wait. In many situations, Trinidadians could be more specific and refer to a universal time or some other time. Sometimes they are more specific; for instance, a taxi driver might say, "when I come again," or someone might say, "Let me do a little thing in the garden." When "jus' now" is used, the aim is to obscure time, not just clock time, and to keep others waiting and ready.

Thus, "jus' now" is used to obscure duration, but in doing so, it emphasizes the event awaited. To say "jus' now I comin'" shifts the emphasis away from the time waiting to the "coming." Likewise, to say "jus' now is Christmas again" de-emphasizes the time between the utterance and Christmas, and it emphasizes the coming of Christmas. In this way, the social relationship is affirmed over and above the time waiting during which the social relationship is suspended. Using "jus' now" suggests an informality that implies solidarity. When used correctly, with the wait being relatively short, it succeeds in emphasizing solidarity, even when its use is prompted by temporal conflict.

To use "jus' now" and then to keep another waiting reflects power or a claim to power over the person kept waiting: "It make you feel very off. But what is it you are going to do? They in authority, so they can tell you 'Wait, jus' now.'"

The use of "jus' now" to demonstrate power over another need not be based upon any form of legitimacy to claim such power, however, and can therefore create resentment. One Monday, a businessman approached a tailor to make a pair of pants. The tailor told the businessman that the pants would be ready on Friday. When Friday arrived, the businessman visited the tailor, who was working on the pants. The tailor looked up and said, "Your pants will be ready jus' now." This angered his client, who pointed out that he paid "good money," and that it was "only professional" to have the work done on time. While, in the conceptualization of class in Trinidad, the businessman is superior and more respected than the tailor, the tailor has temporarily treated the businessman as inferior or equal.

Also, "jus' now" can be used to avoid dealing with another person. An example given by one informant involved going into a government office and having the clerk say, "jus' now." The informant admitted that, in his view, "jus' now" should mean a matter of minutes, but that, since it was a government office, and a government employee was involved, "jus' now" meant a longer period of time:

> I went to [Viego Grande], to the Revenue Office, to inquire about something. One of the fellow tell me, "Jus' now, jus' now." He remain by his table, by his desk, writing, doing this, doing that. He remain there for about a half an hour. He come back and went and tell another fellow.
> That fellow come and say, "What it is?"
> I say, "Well, look, so and so."
> He say, "Well, jus' now." He gone! Where he gone he take another half an hour before he come back to tell me, "Yes, let me

hear what it is now." And when he finish hear what it is, he tell me, "You know, I can't help you there."

This informant then explained that this happened because the clerk "did not care." Other informants used other examples to make a similar point; when "jus' now" results in neglect of the person waiting, it means that the speaker "did not care." Thus, in determining how long "jus' now" means to wait, Trinidadians gauge the degree to which the speaker "cares" about the other person.

"Jus' now" is also used "jus' to be wicked," that is, as a means of deliberately taking advantage of another. In this case, the user of the chat employs it simply to force the intended victim to wait, usually for a long time. This usage is like that of not caring but has an added, aggressive dimension. Since everyone can employ this phrase, it often provides an effective means of expressing hostility to those with more power and authority without suffering adverse consequences. One scenario related to me was about a landowner who asked a laborer to help him on his estate. The laborer told the landowner, "Jus' now, I going to the store to come back"—implying that he needed to buy something at the store. The laborer then limed at the store and kept his employer waiting.

"Not caring" is greeted with disapproval, but it is not the case that every time "jus' now" turns into a long period of time it means that the user of the chat does not care. When one visits a neighbor, and the neighbor is doing some work, the neighbor might say "jus' now," which Trinidadians interpret as meaning "as soon as I am done with the work." Likewise, taxi drivers frequently use the term "jus' now" and mean "when I return from Viego Grande," which can be two hours. When someone who worked for the U.S. Army says "jus' now," they tend to mean a couple of minutes at most. In effect, the meaning of "jus' now" varies according to the situation and the individual speaking, as does time in general. Also, one estimates what "jus' now" means according to the reputation and responsibilities of the person using the phrase.

Trinidadians use expressions such as "jus' now" and "any time is Trinidad time" when the temporal conflict concerns coordination of social interactions, but there are other contexts in which temporal disputes arise. A landowner who hires someone to do work is quite interested in the completion of the task and calculates the wage according to the amount of work. If the wage is a "day wage," as opposed to a "task wage," then, to appear moral, both the employer and the worker must appear to agree that the wage is fair for the amount of work. At the same time, the worker has the motivation to make the landowner feel that he got his money's worth in labor. When the landowner checks with the worker late in the day, the reckoning of when the work was actually finished can have moral and economic implications. If the worker says "I finish long time," it suggests that the worker was quick and efficient. In effect, Trinidadians use the phrase "long time" to evoke temporal distance, even when it does not exist.

Albert, a tailor, always has much work to do. His clients bring him clothes to repair or material from which they want him to make clothes. Because of the volume of Albert's business, he sometimes falls behind the pace of the work assignments he receives. It is inevitable that he has gained a reputation for not always finishing the clothes on time, because, frequently, he does not. Albert is not happy with this fact, or with his reputation. Commonly, customers arriving the day that their clothes are supposed to be ready will reflect their skepticism over Albert's abilities by asking, "Have you finished yet?" Recognizing this subtle attack on his professionalism, Albert, whenever he can, replies, "Long time, long time." In fact, he may have only finished the clothes a few minutes before, even though he seems to assert that the task had been finished for quite some time. "Long time" consigns an event to the distant past but also emphasizes how distantly past the event is. It is contrasted with "a moment ago," or, if talking about periods of greater than a day, "just the other day." "Long time" seeks to exaggerate the amount of time that has actually passed. For instance, students might say they have completed their homework "long time," when they have finished it only a few minutes before. Some informants report that some people say "I finish that long time" before they even begin the task to which they refer.

When used in the context of the completion of a task, "long time" implies an efficiency that sometimes does not exist. Such an assertion, then, is a strategy to enhance the moral character of the people who use the chat. It asserts that they are not lazy, but efficient, and implies that there is no need to inquire about the completion of the task in the first place.

At other times, the use of "long time" severs the present from the past.

This has moral implications that vary according to the context. For instance, I asked an individual when the road through the village had been paved. He replied, "Long time they pave this stretch of road." I also asked the same individual about the insurrection in 1970,[1] and he said it was "jus' the other day." The road was paved in 1978, eight years later than the insurrection. His use of "jus' the other day," which indicates a short time past, and "long time," which indicates a distant past, must involve more than a simple reference to time-reckoning. He does not entertain the idea that the road was paved before the 1970 revolt, but the point that he was trying to make with the use of these temporal terms was not based upon locating events in a chronological past, but in explaining the past in relationship to the present.

Upon being asked for his justification for saying that the road had been paved "long time," he pointed out that it has never been resurfaced, and that it was in deplorable condition. The present poor condition of the road formed the basis of his description of its surfacing as being a distantly past event. The original surface had shown significant wear after twelve years and badly needed resurfacing. When I asked about the 1970 insurrection being "jus' the other day," he said that the recent attempted coup[2] had reminded him of the previous insurrection. Like many Trinidadians, he saw a relationship between the previous revolt and the more recent attempted coup. Therefore, the use of these terms to characterize past events actually establishes their relationship with the present.

Not only is "long time" used to establish a history of neglect, but its other use with regard to relating so-called historical events is to establish a discontinuity between the past and the present. For instance, the way in which Anamatians cared for the recreation ground "long time" is very different from the present. "Long time," Anamatians maintained the ground themselves and cut the grass by hand, but now they rely upon the County Council to cut the grass with a tractor. The use of "long time" in this context emphasizes how differently people acted in the past. By doing so, it prevents the integration of past and present, thus positing a temporal discontinuity. Therefore, its use is to emphasize the present. Even when people refer to how things were "long time," it is done with the assertion that things are no longer like that and that the present reference to "long time" is didactic.

Glosses differ from tokens in that they obscure temporal realities and temporal differences rather than present a facade of agreement. They serve a different purpose, being deployed where temporal agreement or an illusion of temporal agreement is not as important as coordinating social relationships or making a moral point. "Jus' now" and "long time" are not matters to be agreed upon, but phrases to be accepted and reacted to in socially appropriate ways.

### Changing Times

The final strategy for dealing with temporal conflict is to change a definition of time so as to eliminate difference and generate solidarity. Its use involves a series of complex issues concerning solidarity and power. In this chapter, I shall concentrate only on the use of changing times in the pursuit of maintaining solidarity in a situation initially marked by temporal difference. The use of this strategy to enhance group exclusivity will be discussed in the next chapter.

The strategy of changing time is to eliminate temporal differences. A 1991 protest about the condition of Anamat's roads provides an example of a large number of people changing their times to express solidarity. One morning, a small group of Anamatians decided to protest the poor condition of the road by creating roadblocks. By blocking the road, the major artery out of Anamat, they ran the risk of alienating a large proportion of the village's population, but since it was not a market day or the beginning of the month when the government checks are released, the actual disruption was minimized. It affected primarily taxi drivers and their passengers—mostly schoolchildren attending school outside the village. The morning of the protest, after the overseer for the Ministry of Public Works and Infrastructure had gone to the Anamat office, the organizers blocked the road behind him, effectively preventing him from leaving. Soon after, other people began to suspend their work in order to participate in the protest. A crowd gathered and fortified the first road-blocks further and made more roadblocks. By suspending their work, they suspended their work rhythms. Farmers disrupted their daily routine of going to the fields, and many taxi drivers parked in front of the road block, choosing not to make further trips back to Viego Grande, even though they could have worked the road from the road block to Viego Grande effectively. Those who normally watched soap operas from noon until two forsook this daily ritual in order to become a part of the crowd. For the group that gathered, almost every participant, except for the boys from the block, had to deviate from their normal diurnal rhythms and associated times. Their willingness to do so showed support for the protest and solidarity with the other villagers.

Changing time can also reflect formality. If a person makes an appointment, and then keeps the appointment because of its importance, then such a case of switching occurs. Trinidadians associate clock time, scheduling, and appointments with respectability. A group of individuals who wish to make a public performance of respectability to one another will schedule their events according to clock time and arrive punctually. By doing so, they communicate to each other that they are respectable and view each other as respectable.

What I call hyper-punctuality is an example of such formality. Hyper-punctuality is an effort to be perfectly punctual, whatever the appropriate punctuality is. Frequently, it is a clumsy attempt to adapt to a temporal system to which one is unaccustomed. When I would appear at a fete exactly two hours late, I did so consciously, and its exactness not only reflected my unfamiliarity with the temporal system of fetes, but it also expressed a sincere effort to arrive at the appropriate time. Likewise, when Trinidadians who were supposed to visit me at a particular time appeared exactly at that time, or just a minute or two late apologizing profusely, it revealed the same unease. In such cases, the informant would not realize that, in the model of time that I employ and share with many American academics, tardiness of only a minute is not to be construed as tardiness at all, but as being punctual. In these cases, individuals attempt to adopt another temporal model and formalize it so that it is even more rigid than what is normally applied.

### Conclusion

This chapter has discussed how solidarity is maintained in the face of temporal difference. Since formal institutions do not extend their temporal definitions into most of everyday life, temporal conflict frequently arises in the unfolding of social relations. This is because social relations are coordinated by a variety of different times, and each time has its own form of punctuality. The way persons conduct themselves when confronted with temporal differences reflects power, solidarity, and structural conflicts, such as those between ethnic groups and classes. Temporal differences are managed in three ways: tokens, glosses, and changing times. Tokens postulate temporal agreement where there is none. The tokens employed by Trinidadians include "any time is Trinidad time" and "time is time." Interestingly, these tokens refer to opposing positions, and the fact that the same individual will use both reflects their quality as tokens. Glosses obscure temporal difference and, instead of requiring agreement, are used to elicit a response from their audience. Because they have a direct influence on how people act, they are very effective in coordinating people who hold different ideas of time. Finally, changing times involves an individual switching the conception of time he or she applies in a situation in order to be consistent with others.

# Social Conflict and Time

The population of Anamat, like that in the rest of Trinidad, is very diverse—age, gender, class, and ethnicity all define the diversity of Anamat. When interpreting interpersonal conflicts, Anamatians use these categories to discuss the origin of the conflict and to determine whom they support. In cases of ongoing tension, the public performance of difference and antagonism between members of different social categories frequently involves the use of ideas about time. Since, in Trinidad, relationships are intricately coordinated in time, expressions of antagonism through the medium of temporal dispute can have serious repercussions: the disruption of the coordination of one social relationship often leads to other relationships being disrupted as well.

## Interpersonal Temporal Conflict

In Anamat, automobile repair is one situation that often creates temporal conflict. Repairmen take work on a first-come-first-served basis. Unless the repair is complex and will take a long time, the owner of the car remains at the shop, watching the work and talking with the mechanic. Consequently, most car owners have clear conceptions of how much time various repairs take, because they witness them being performed. The car owner judges whether to stay or to go by considering the mechanic's initial diagnosis, as well as the number of other jobs that the mechanic must complete before beginning the new repair. Typically, then, when one takes a car to the mechanic, one has an expectation of when it should be repaired, depending upon the complexity of the problem. The car owner sees a relationship between the amount of labor required and the amount of time the car should spend in the repair shop. When a mechanic does not share this relationship between time and completion of the work, problems erupt.

One such case arose between a young Creole auto-body repairman and an Indian shopkeeper in his mid-thirties. The repairman was known locally for performing exceptionally attractive and flashy body work. He was also suspected of being unreliable—a rumor that drove a great deal of business away from him. The Indian shopkeeper was this repairman's friend but had never been the man's client until he needed some body work performed on his van. He decided to take it to his friend for the work to be done. The shopkeeper could not afford to lose the use of his van for very long: since Anamat is a remote village, only a few types of goods are delivered to the local shops, and no staples like rice or flour are delivered; for the most part, shopkeepers must drive to Viego Grande in order to purchase goods from wholesalers, so having access to a functional vehicle is a prerequisite to keeping the shelves properly stocked.

Repairmen do not have to be concerned with other individuals' rhythm of activities. The popular repairmen are concerned out of courtesy, but the auto-body repairman that the shopkeeper used did not fall into this category. This repairman has come to view his work as more than simply repairing vehicles. He views it as art and endeavors to incorporate his own flamboyant style into the vehicles of all the customers who want the exteriors of their cars and trucks repaired and painted. In addition, he tends to work only when inspired, not necessarily when required. In sum, this man values his work according to his estimation of the elegance of the final product, and he feels that his clients should do likewise.

When the two men discussed the job, both agreed that it would probably take several weeks for completion. After several weeks, the owner became anxious about the completion of the job, and when he checked on its progress, he discovered that the repairs were far from complete and that flamboyant modifications had been made that were not part of the original agreement. Aggravating matters, he frequently saw the repairman driving his car around and liming. After several more weeks, full of confrontations, the owner physically threatened his former friend, seized his unfinished van, and took it elsewhere. The repairman complained that he had put more money and material into the van than for what the shopkeeper had paid him and that the shopkeeper did not appreciate quality. The owner argued that he had paid more for the work and materials than was originally agreed upon and that when a job needed to be done promptly, he expected it to be completed.

As the dispute continued to grow, both persons attempted to garner moral support from other villagers, basing their arguments upon two different sorts of claims. First, they disputed what constitutes a fair price for the work and cost overruns. This strategy had little popular appeal. Most observers noted that they had not been privy to the initial negotiations and that they could not make a judgment of this matter. The second argument was based upon time and deprivation. The owner of the vehicle required it for his job and had anticipated being without it for only a couple of weeks, but after six weeks had passed and his vehicle was still in pieces, he began to complain. His complaint about how long the job should take met with support, particularly among taxi drivers, marketers who sell produce in Port of Spain, and other shopkeepers, all of whom also depend upon their vehicles for their livelihood and understand what is involved in repairs. Unfortunately for the reputation of the repairman, shopkeepers and taxi drivers play pivotal roles in the passing of information in Anamat—both talk with their clients about the latest news and gossip. The repairman's appeal to craftsmanship did not satisfy these multiracial groups that shared in common their interests in practical matters. However, the repairman did gain support from two groups: his immediate family and a few young adult Creole men. The former group was based upon kin solidarity. The latter group was based upon ethnic, class, and age solidarity. They felt that the owner of the vehicle was a miserly Indian who, in their minds, had cheated Creoles, and therefore he deserved to be cheated by a Creole. The former did not address the issue of time. The latter referred to the vehicle owner's ethnicity and upward mobility and saw him as too impatient, too concerned with time, and too miserly.

This is a case of two different associations between time and work conflicting. The mechanic did not incorporate a bounded conception of time into his conception of the task, whereas the owner of the vehicle did—largely because of the time constraints on his occupation. Each disputant appealed to their ideas of time and work for moral support from the community.

In this case, time and temporal conflict involve claims of power and authority in disputes. Indeed, time is a resource for the exercise of power to which almost everyone has access. Through the use of temporal conceptions, power can be exercised without the expenditure of any material resources. In the case of the dispute between the shopkeeper and the repairman, claims of legitimacy in the dispute rested, in part, on temporal claims.

The shopkeeper and the repairman did not base their claims in their dispute on ethnic or class differences. The shopkeeper occasionally talked in terms of the need for the repairman to "grow up" and to "stop liming all about," but this was an appeal to maturity and age rather than to ethnic differences. Despite confining their argument to issues of personal respect, many Anamatians applied ethnic stereotypes in interpreting the dispute, suggesting that the Indian shopkeeper was miserly or that the Creole repairman had no respect for time. In this way, the conflict came to reinforce these stereotypes even though, when it began, ethnic difference was not a major component of the dispute. What began as an interpersonal conflict came to reinforce discourses of ethnic difference.

In many ways, the use of time differences in personal conflicts between those of the same ethnicity, age, class, and gender resembles its use in structural conflicts between ethnic-, age-, class-, and gender-based groups. In both personal conflict and structural conflicts, Trinidadians tend to relate issues of time to claims that establish some form of superiority, whether in terms of logic, morality, or power. In structural conflict, Trinidadians use times to express loyalty to one group and difference from another. In personal conflict these structural issues frequently emerge, even though initially the appeal is to individual respectability and moral standing rather than to structural differences such as ethnicity.

In addition to expressing ethnic, gender, class, and age conflicts, ideas of time express and maintain conflict between other social groups, such as kin groups and residentially based social groups. In Anamat, there has been a long-term rivalry between those who live "up the road" and those who live "down the road." For the most part, different kin groups live in each area, but the ethnic and class composition of both areas is mixed.

These social differences in Anamat are quite salient and have played major roles in village politics. Since the Village Council serves as a conduit for requests for benefits from the national government (Craig 1985), the Village Council has, at various times, been a forum of competition for resources between the "up the road" residents and the "down the road" residents. In open conflicts, the use of time has served as an important political resource to delay or disrupt meetings and, consequently, decisions. The ability to do so has been based, in part, on the Village Council having to work within the temporal limits given to it by the national and county administrations.

## Class Conflict

In Anamat, class differences become displayed through sparring and concessions over time. The way in which Anamatians publicly deploy ideas of time expresses solidarity, power, and resistance to that power. By asserting a place within the matrix of these three dimensions, individuals situate themselves in relationship to others in terms of powerful categories of class, age, ethnicity, and gender.

Differences in times have been referenced frequently in the historical study of the Industrial Revolution, as well as in more recent studies of class conflict (Thompson 1967; Willis 1977). These discussions argue that the class that benefits from production attempts to assert a particular definition of time upon laborers in order to maximize production but that the workers resist with their own, subversive ideas of time. With regard to the working class in England, Willis (1977) provides examples of how antagonism to the form of time defined by industry and taught in school begins during children's education and then is incorporated into their habits as laborers. I suggest that, wherever members of different classes meet, they may use time as a means of expressing solidarity with their class and antagonism toward those they perceive to be members of a different class.

In various contexts, adopting a particular time can be an act of domination or a reaction to domination. In Trinidad, where ethnic and class categories are not always clearly distinguishable, when two people meet, they are likely to negotiate their relationship as they discover exactly who the other individual is with regard to ethnicity and class. Since time is a major means of publicly delineating such differences, the use of temporal ideas can change within a single social setting. In Anamat, this does not happen often, since all the residents know one another, and complete strangers visit rarely. But Anamatians frequently must deal with strangers, and particularly with strangers in positions of authority outside the village.

The doctor-patient relationship is one such case of an encounter between strangers. In Trinidad, medical doctors vary with regard to how they manage their patient loads. A few doctors choose to arrange appointments; others prefer to see their patients in the order of their arrival. These two different strategies involve two different models of time. The former links social relationships to particular temporal periods as defined by a clock, whereas the latter follows a task orientation, with priority being given to the order in which patients appear rather than to when they have scheduled an appointment. Both means of coordinating numbers of patients can be disrupted by emergency cases that demand immediate attention.

Generally, doctors who schedule appointments rather than see patients as they come are either specialists or they see a clientele that is not drawn from the lower class. Since Trinidad has a system of socialized medicine combined with physicians in private practice, a doctor who charges fees higher than other private physicians will attract clients who can afford to pay the high fees. Trinidadians regard accomplishing this level of practice as prestigious, but it is often accompanied by behavior that asserts the respectability and superiority of the physician in their relationships with patients. Some doctors choose not to respect their schedules and not to adopt a first-come-first-served treatment of patients but instead treat appointments with favoritism according to ethnicity or class, or they might even choose to disrupt their patient schedule with social engagements, such as lengthy lunches. Such individuals construct a schedule based upon clock time and ask their clients to adhere to the schedule, because doing so is respectable, but then the doctors disregard the clock-defined schedule with respect to those whom they consider less respectable and, by implication, of a lower class.

Victims of this sort of treatment interpret the doctor's behavior as arrogant and condescending. They complain that such doctors "don't care," even though they are in a position in which they are paid to care. Trinidadians view not caring as an assertion of superiority. When a youngster makes such an assertion, then that youngster can be reprimanded or disciplined, but a doctor or a government official is beyond effective reproof from most visitors. Those who feel the bite of this condescension react by challenging authority: if the doctor will not be in the office in time for the appointment, then there is no need for the patient to appear at that time, either. Consequently, the patient arrives late and the doctor arrives late. If the doctor arrives before the patient, it becomes fodder for resentment expressed through even greater condescension, even with the doctor going so far as to suggest that a patient's being late for an appointment forfeits his or her right to see the doctor any time soon. If the patient arrives before the doctor, even though the patient arrives late, then it also fosters resentment in the patient. Once this sort of relationship is put in place, the complementarity of condescension and reaction defines the ongoing relationship.

Frequently, these sorts of relationships take place between members of different classes. Since the most frequent contact between those of lower class and upper class is in such formal settings, these settings become the field of battle for continuing temporal disputes that perpetuate themselves. An appointment is set, and then both groups miss the appointment—one does so as a statement of power and authority, while the other does so as an attempt to subvert that power.

What the examples of doctor-patient relationships show is that, in relationships in which different social categories meet, the use of time becomes an important dimension for exhibiting differences. In a sense, in any situation there are three possibilities: first, the persons involved might employ complementary times; second, they might use a chat to gloss over temporal differences; or third, they might employ conflicting times and decide not to gloss over the differences. In some situations, such as the appointment with the doctor, the time deployed might be in conflict both with the situation and with the other individual in the situation. When this occurs—a deliberate attempt to generate conflict—it is usually because of a larger rift, such as class. In the example of the doctor, the expectations of the doctor are very clear, and the power that he or she has over a patient is also well understood by Trinidadians. Consequently, the sorts of temporal behavior one can expect are also easily anticipated. Complications arise when individuals are equal in terms of power but are different in other respects, such as between members of different ethnic groups.

Situations in which different classes meet are not always occasions in which the higher-class member asserts power and domination over others. In many cases, members of different self-defined classes come into contact in business relationships. In these relationships, the lower-class individual disrupts the other individual's time, thereby undermining that person's authority. This is the case in many service industries. Returning to the example of tailors discussed in the previous chapter, tailors have daily opportunities to do this. For small projects, like alterations or even making a pair of trousers, most tailors know that they can finish the project quickly. A typical scenario involves someone who perceives himself or herself to be of a higher social stratum asking the tailor to undertake such a minor project and giving the tailor a specific time frame in which to finish, in effect expressing power over the tailor by defining the tailor's schedule. The tailor dutifully takes the cloth or the item to be altered, puts it on top of a pile, and makes a comment such as, "No problem." The client returns at the appointed time and asks for the article of clothing. The tailor replies, "Jus' now," but the client does not find this to be a satisfactory answer. The client usually replies by saying something like, "I told you to have it done by this time today, and I meant it." In many tailoring shops, there are no clocks, and the tailors can credibly reply that they do not know the time. The client meets such a gambit with the suggestion that it is indeed the time for the project to be finished, thus eliciting the response, "I must have lost track of the time." The tailor then puts down the project on which he is working, picks up the project for the now-irate client, and informs the client to "come back in a while." The tailor then works on the project, having it close to completion when the client returns.

When this scenario occurs, as it does quite commonly, the client has only nominal recourse against the tailor. He or she can take the project elsewhere, but, typically, he or she will receive the same treatment elsewhere. To react with too much ire makes matters worse—for example, the tailor might not be finished when the client returns and will make the client suffer the indignity of sitting and waiting in the tailor's shop while the task is finished.

It is not the case at all that tailors have no sense of time. Indeed, among all those interviewed, tailors articulated a conception of time as a commodity more than other occupational groups. They were one of the few groups who used the expression "time is money" and pointed out that, in their profession, success is evaluated in terms of quality work performed quickly. Tailors are also acutely aware of time passing. While many do not wear watches or have clocks in their shops, they often listen to the radio and are adept at judging the time according to the regular news broadcasts and the shift changes of disc jockeys. While through interviews, tailors appeared to be the group most conscious of time as commodity, when nontailors were asked about tailors, a very different image emerged. Tailors have the reputation of being unreliable with deadlines and punctuality. Clients who are least willing to wait in a tailor's shop have the greatest difficulty. Clients with the fewest problems are willing not only to wait for their work but also to lime in tailor shops.[1] The greater the expression of equality with the tailor, the greater the likelihood that the tailor will complete one's jobs punctually; the greater the expression of condescension through the presumption of controlling the tailor's time, the greater the difficulty with the tailor's punctuality, or lack thereof.

The use of temporal conceptions, the deployment of times, and contesting the imposition of ideas of time by others all serve as means of expressing and maintaining class differences. Since, as mentioned previously, time is a resource available to most people, even the lowest classes, it serves as a means of disrupting the lives of others, even those with far more power and material resources than oneself. Consequently, in many situations in which members of different classes come into contact, power is displayed through manipulation of others' time, both in terms of domination and of resistance to domination.

### Ethnic Conflict

Often, in Trinidad, class conflict converges with ethnic conflict in subtle ways. In situations where members of different classes are also members of different ethnic categories, they can include temporal stereotypes as an additional resource in defining and contesting power. As a dimension of difference, ethnicity is far more marked than class in Trinidad (Singh

1994). Trinidadian party politics reflect ethnic differences and tensions (Hintzen 1989; Ryan 1972). Currently, the political party that makes the most explicit appeals to class, the National Joint Action Committee, is unable to muster much popular support, whereas political parties that have based their appeal on ethnic affiliation have been quite successful in winning seats in parliament (see Hintzen 1989). To a great extent, ethnic differences obscure class differences in Trinidad (Singh 1994:xiii–xiv).

Temporal stereotypes are associated with the two largest ethnic groups in Trinidad, the Indians and Creoles. These stereotypes emerged out of the interaction of labor and ethnic relations. Creoles are described as present-oriented and interested only in flamboyance and conspicuous consumption, and Indians are stereotyped as future-oriented and interested in using time to make money. An Indian man commented,

> The Indian, they didn't waste time. They work, and they provide, but the Creole people now, they always have a plan, but when the money does come, the plan does stop. Plenty of them had estate, and is the same thing, they come and they lost it. Too much of fete and they mortgage it. So the Chinee or the white man end up with it. So, plenty of them have land, and they have to lost it through that they waste their time. But, the Indians, they didn't waste their time. Only Indians will make a proper day work because they studying for the future.

The validity of these stereotypes has been disputed, however—there is evidence that Creoles are as future-oriented as Indians (Baksh 1979).

Indians are seen to be thrifty with their time and their money; Creoles are stereotyped as paying little attention to the value of either. These stereotypes are mirror images of one another and form the basis for interpreting much behavior with regard to economic success. Many Trinidadians argue that Indians tend to be more successful at small business enterprises than Creoles because their attitudes toward time and money are the foundation for business acumen. Others argue that Creoles are not successful at such businesses and tend toward wage labor because they are not inclined to save and therefore cannot succeed in business. The reality is that, while Indians dominate some sectors of the economy and Creoles dominate others, this is a result of networking and nepotism, rather than any particular business skill (Ryan 1991b). Furthermore, these patterns are the product of a 150-year history of the Trinidad elite's differing treatment of Indians and Creoles in law (Samaroo 1985; Singh 1985, 1994), labor practices (Brereton 1979, 1981; Singh 1985, 1994; Wood 1968), and ideological representations (Segal 1993; Singh 1985). Indians and Creoles have developed several divergent practices with regard to kinship, marriage, and the cycle of domestic groups, as a result of this history. These differences influence Creole and Indian ways of planning for the future, but not in the sense of present-orientation versus future-orientation.

Indian families act as corporate groups, but Creoles tend to act as individuals and use their kin as resources for individual endeavors, not as partners for corporate endeavors. When one discusses the future with Creoles and Indians, one notices something very different. Creoles tend to plan their future in terms of individual success, whereas Indians tend to view their future in terms of success building upon a previous generation. Indians commonly plan for a future, for the benefit of their family. As one Indian man said, "I want to expect something better for my family, for the home, for myself." This is in contrast to a Creole man who reported: "Right now, my plans are like when I go up there [to the U.S.] and work, after that come back here and work and look to settle myself: get a pickup and work, employ myself, be a self-employed person, work on me own. That is how I think I have the future set off."

But these differences are simply tendencies, and in fact they are not determined by ethnicity, per se, but by the nature of kinship links within Anamat. While it is more common for a kin group to be a corporate group among Indians, there are some Creole corporate kin groups, and in these cases, the individuals also emphasize the family in their discussions of the future. The reason for these kinship differences has to do with a powerful, senior member of the kin group controlling an important resource, such as land, and coordinating the activities of kin in the utilization of the resource. In Creole kin groups, the use and control of land tends to pass from mother to son, and from husband to wife. Among Indians, such use and control is patriarchal, passing from father to son. The patrilineal inheritance pattern among Indians is consistent with the patrifilial tendencies of their kin groups and provides for the foundation of the kin group's corporateness. The inheritance pattern among Creoles generates rifts within the kin networks, making it difficult to maintain a corporate group across several generations. This, in turn, fosters greater independence from the set of kin as a whole, but because of the importance of mothers, it fosters greater ties to the mother and to one's maternal siblings. The result is that Indians' plans are embedded in their families' future plans more often than Creoles' plans are embedded in their families' future plans.

With regard to the stereotypes of Creoles and Indians, it is not possible to identify particular times as definitively Creole or Indian. Indeed, these different stereotypes can be applied to the same behavior. In discussing Creole government workers, Indians commonly attributed the lack of "caring" and "punctuality" on the part of these workers to the stereotypic notion that Creoles have little sense of time. On the other hand, when interpreting the lack of "caring" and "punctuality" among Indian businessmen, Creoles attributed this to the stereotypic Indian attributes of not being concerned about others' time, but only about their own wealth. Within these stereotypic representations, if a Creole arrives late for an appointment and states that "any time is Trinidad time," it is viewed as an example of the Creole lack of a sense of time; if an Indian arrives late for an appointment and states that "any time is Trinidad time," it is assumed that the Indian had something more important to do than the appointment, and it is thus assumed to be a sign of Indian arrogance and lack of caring. The behavior is the same, but the stereotypes used to interpret the behavior differ and are determined by the ethnicity of the person involved.

The fact that there is little validity to the stereotypes does not undermine their widespread use. Stereotypes serve as useful resources in the maintenance of social divisions for political purposes. Anamatians use the stereotypes as a resource in disputes. They use stereotypes reflexively to form a bond of solidarity with other members of one's ethnic group, or they deploy stereotypes aggressively to attempt to insult a member of another ethnic group.

For instance, one Indian head of a household who frequently hires agricultural laborers commented that Creoles are too lazy and do not know the value of money. He explained that they want to get paid for little or no work, and then they get angry when they are forced to do work. His basis of comparison was between young adult Creole men and himself and his sons. He argued that Creole men did not work as hard for the money they were paid. In effect, their labor was not worth wages any higher than he paid. Implicit in his reasoning was a recognition that

Creole men did assign their time a monetary value, but that the value they assigned was higher than the value he assigned. His rhetoric was that Creole men had no sense of time, but his argument implied that they were acutely aware of time as a commodity.

In a different context, one Indian villager tried to show that Indians can fete and spend money as freely as Creoles. This form of conspicuous consumption and apparent disregard for time forms part of the stereotype held of Creoles. He felt unjustly accused of being miserly and wanted to show how generous he was, and how he could disregard his business interests. The Creoles present angrily refused his overtures and explained that he was just trying to act "big," and that his offer of free drinks did not fool anyone—they asserted that he was still a miserly, proud Indian who insulted Creoles regularly. In fact, the Creoles interpreted his intentions as utilizing a fete context to generate more business for himself later. In Anamat, one's choice of shop is influenced by one's social relationship with the shopkeeper, not just by the prices charged at the shop. Thus, the social overtures of a shopkeeper can readily be interpreted, rightly or wrongly, as having a concealed business motivation. In this particular case, Creoles perceived that the shopkeeper was bringing business into a fete situation, and doing this undermined the ideology of fetes being contexts in which no commitments, even informal ones, arise. The Creoles then suggested that taking the time at a fete and using it to generate business fit the stereotype of Indians using all time to their own miserly ends. Consequently, whatever the Indian man's actual motivation was for his actions, his actions were interpreted with stereotypes of Indians and their use of time, and then used by Creoles as a means to maintain a social distance.

Anamatians also use these stereotypes to reinforce ethnic self-consciousness and cohesion. Creoles will say about themselves that they "fete too much and lime too much." Such a statement is generally made while Creoles are working or trying to find work, or when they have just finished working. In such a context, its tone seems negative and even fatalistic, yet it is almost always associated with two other issues: the differences between Creoles and Indians, and an optimism concerning future economic prospects. With regard to the latter, the juxtaposition of a self-imposed stereotype of poor management of time as a commodity, and a "now for now" mentality, with a discussion of future plans seems contradictory. Yet, as long as the discussion of Creole time-consciousness is explicitly or implicitly linked with Indian time-consciousness, it is a subtle articulation of the economic success of Creoles not being based on miserliness, in contrast with the stereotype of Indians.

Indians use the stereotypes in a similar way. In contexts where they are feting and conspicuously consuming, they do so as a performance of their greater perceived ability to use time to their advantage. On these occasions, they out-spend and out-consume their Creole counterparts. Such occasions, when organized by Indians, such as weddings, bazaars, and fetes, are presented as the fruits of Indian hard work.

Thus, the implication of these temporal ethnic stereotypes shifts from context to context. They serve as tokens to unite one group against another. In some situations, articulated, publicly recognized differences provide an excuse for conflict. As a result, personal animosity takes on the dimensions of class and ethnic identity when people or groups seek support for themselves. Commonly, in a conflict between an Indian and a Creole, those directly involved attempt to mobilize support on ethnic grounds by applying derogatory terms and stereotypes to one another. As was shown previously, many of the labor disputes in the village take this form. In addition, these stereotypes become incorporated in interpretations of societal conflicts.

## Age Conflicts

As was argued in chapter 2, the relationship between age cohorts is as important to Trinidadian social structure as is ethnicity. While Trinidad does not have a formal age set or age grade system, older age cohorts control resources and wield power. This creates tensions with younger age cohorts. Again, the use of time establishes, performs, and maintains such conflicts. One of the significant differences between youths and old heads is in the value orientations that Peter J. Wilson (1969) labeled as "respectability" and "reputation." Respectability is associated with old heads and women, and reputation is associated with young adult men. As individuals move through the life cycle, they move from the value orientation of reputation to the value orientation of respectability. The ways in which people express these value orientations are multiple and various, including sexual behavior, attitudes toward marriage and child rearing, and economic behavior. Those who present themselves as respectable tend to utilize the token "time is time" to talk about the relationship of work and time. As indicated in chapter 3, this relationship varies considerably. In addition, as discussed in chapter 2, each generation's conception of the relationship between time and work differs. Since older generations tend to hire or supervise members of younger generations, these differences become important.

Consequently, each age cohort has, to some extent, experienced very different conceptions of the link between time, labor, and wages. When members of an older age cohort employ members of a younger one, the disputes that come about reflect these differences. Most often, the conflict is between an employer who worked for expatriate managers and young workers who previously worked for Trinidadian managers in the public service. The young worker realizes that he or she is likely to be paid only for the time worked, and the worker expands the tasks required so that they result in one day's wages. The old head employer wishes to pay for hard work and attempts to prevent workers from slowing their pace to make more money by extending the work over more days than required. Debates arise over the total wages, and many old heads choose to avoid such confrontations by paying piece-rate whenever possible. But even here, there is debate over the appropriate wage per task. Both employers and employees agree that piecework avoids disputes over the relationship between time and wages, although it retains disputes between work and wages. The major benefit of piecework over hourly wage work for Anamatian employers and workers is that in piecework agreements, the debate over wages is resolved before the work commences. If no agreement is reached, then no work is performed. In the case of hourly wages, there are often disputes after the work is completed. Employers complain that the amount of work done in the time worked is too little, and workers complain that the employers pay too little for the amount of work done in an hour.

Possibly, the present old heads did the same when they were young but have since adopted the conception of the time that was being forced upon them in their first work situations. The evidence for this comes from individuals who are middle-aged. Interestingly, this cohort does not have a distinct label, as to the youths and old heads. Also, this cohort exhibits a substantial shift in behavior from a value orientation toward reputation to one of respectability. Reputation emphasizes behaviors that elicit public recognition. For men, this includes conspicuous spending of money and demonstrations of virility through having multiple lovers and fathering children (Wilson 1969:73). Respectability emphasizes behaviors that conform to ideals of sexual propriety and social responsibility promoted through the moral teachings of the Christian church and the schools (Wilson 1969:77–78; 1973:229). Over the course of the twenty continuous months I spent in Anamat during my first field trip, several men between thirty and forty years of age exhibited this shift in value orientation, at least in the opinions they expressed when they limed with other men. Several reported that the growing responsibility for their children prompted their value reorientation. One referred to a fight in which he was badly injured as a clear signal that he needed to changes his ways. They all moved away from a discourse of reputation, which involved bragging about sexual conquest, liming on the block, gambling, and hustling, to a discourse of respectability, involving faithfulness to one's wife,[2] avoiding drugs, controlling one's drinking, and working hard to support one's family. Likewise, before the attitude shift, these men did not easily conform to any definitions of time thrust upon them, but afterwards, they were extremely industrious, taking a variety of jobs and conforming to the ideas of time embedded in the job. An example of this is Arnold. Arnold had been economically successful but had squandered his money on women and gambling. His wife left him, and he moved back to Anamat to live with his parents. As his children reached adolescence, Arnold began to "study" how his son was making the same mistakes he had. When Arnold was injured in a fight one night, he decided that he had to change his life. After he recovered, he left Anamat again and went back to work to earn money for his son, to set a good example, and to become respectable. The orientation toward the connection made between time and work, then, is closely tied with value orientations.

These conflicts, then, become material through which the divisions between age cohorts are defined. With age and the value orientation of respectability, workers tend to attempt to please their employer more than previously. When an individual makes this shift, which involves deferring the ability to define the relationship between time, wages, and labor to the employer, the individual begins to be viewed by the other villagers as entering an unmarked category of middle age. With this comes the ability to assume minor leadership roles. With no formal markers of age sets or age grades, such as rituals, public performances of appropriate behavior for the age cohort become an important component of marking changes throughout the life cycle. With such changes, and the public performance of the markers associated with such changes, conflicts ensue between age cohorts. Minor temporal disputes arise between middle-aged individuals and their young adult former peers. Sometimes, the middle-aged person begins to lecture his or her former peers on the virtues of hard work and respecting time. This elicits a response among the young adults, in which they reflect on how recently the newly respectable individual was just like they are, and such reflection often includes elements of hyperbole, which portray the newly respectable individual as having been even worse than the young adults he or she is criticizing.

In this fashion, the deployment of conceptions of time and a discourse on time in public settings mark the passage of an individual from young adulthood to middle-aged adulthood. These times and discourses are deployed in such a way as to prompt minor disputes and conflict. Typically, older heads suggest that the youths have no respect, and that "time is time."

Because of this connection between age and ideas of time, the expression and use of temporal difference becomes one means of defining oneself with regard to age in relationship to others. Age difference also serves as a means of justifying a conflict by using temporal means. The old heads view the youths as lacking temporal discipline, and the youths see the old heads as manifesting a temporal discipline that is sometimes arbitrary. When the youths choose not to listen to the old heads—that is, when they defy their authority, the old heads often attribute the problem to a lack of temporal discipline.

### Gender Conflict

There are no times that anyone must conform to because of gender. Because there are roles associated with particular genders, and because time is closely tied to such statuses, temporal differences occasionally become incorporated into larger gender conflicts. Temporal differences serve as one of many dimensions along which disputes between men and women arise. Also, the relationship between male heads of household and members of the household tends to be a relationship defined, in part, by power expressed through the control of time. The expression of domination by a male household head and the female household head's need to coordinate all the relationships within the household result in a temporal collision. It may seem counterintuitive to have two heads of household, but the dynamics of matrifocality lend themselves to this competition for power in the household. In this system, a man is responsible for providing for his spouse and children economically. Limited economic opportunities for the men result in their conjugal partners making economic decisions for the family, often in consultation with the oldest children (R. T. Smith 1996:42). This causes a situation where there is a "combination of an expectation of strong male dominance in the marital relationship and as head of the household coupled with a reality in which mother-child relations are strongly solidary and groups of women, daughters, and daughters' children emerge to provide a basis of continuity and security" (R. T. Smith 1996:45).

Time forms a dimension of conflict between men and women in this context. Such conflict is primarily manifested in the disruption of rhythms—particularly those rhythms exclusively associated with the opposite sex. Consequently, men attempt to disrupt the women's rhythm of watching soap operas, and women disrupt men's rhythms of meals and, when given the chance, limes, as well. For elaborate limes, men ask women in the household to cook large meals to take—the food is taken, and the women are left at home. For instance, every year, Randall organizes a lime to go to the preliminary round of the Panorama steelband competition. Each year, he tells his "partners" that they need to get an early start, and that his wife, Elizabeth, will cook food to take. It turns out that Randall's ability to make an early start depends upon his wife's getting up early in the morning in order to prepare a meal. Every year the same pattern is repeated, however: Elizabeth does not have all the ingredients she needs, so she does not get up early to start cooking; she sends Randall to get the ingredients; she then starts cooking late; and the lime begins later than Randall wished. Even on those occasions when Elizabeth has all the ingredients she needs, she takes her time cooking. In effect, on such limes the meal gets prepared, but not within the temporal parameters desired by the males. When the lime finally leaves, Randall's partners complain jokingly about the late start, and, in reply, Randall blames the delay on Elizabeth. When she discusses these limes with other women, they acknowledge that the delay in such limes is their fault, but they add that if their husbands had provided them with the ingredients for the meal early enough, then there would have been no delay. The women also suggest that if they were invited on such limes, they would be more inclined to start cooking early than when the men leave them at home. In men's opinion, if women come along, they disrupt the lime, and if women have to prepare food for it, they delay it. Men suggest that women disrupt limes by requesting to stop at places the men do not want to stop, and by preventing men from stopping and liming at rum shops at which they do want to stop. Women recognize their disruptive qualities and capitalize on them. They have the ability to disrupt limes that they are left out of and often even resent, since limes can sometimes serve as occasions to meet mistresses.

Some men verbally recognize that they make unreasonable demands upon women, and that they should take their wives on limes more often. To a large extent, this serves as a token to placate women, because women very rarely get taken on limes. Instead, men continue to attempt to exercise power over the women by demanding that women cook for events in which they cannot participate. As in so many other cases of temporal conflict, the different sides encourage one another. The delaying of limes is a means of subverting men's exercise of power to define time, and the male attempt to define women's time is a response to the women's subversion. What emerges is an endless cycle of assertion and subversion.

The women react to male attempts to define time by carrying out cooking tasks according to their schedule, not because of men's orders, and through watching soap operas. Indeed, every male attempt to define and structure time for women is met, at some point, with some reaction. When men demand that women cook fresh meals, rather than heating food previously cooked and stored in the refrigerator, the women respond by preparing the meal at their leisure, or by serving "fridge food" without the men knowing, or by preparing major components of the meal and refrigerating these components before cooking, or by preparing fresh, hot food and allowing it to sit on the stove top for several hours until it is reheated or served lukewarm. There is only one exception, and that is found in the power mothers exercise over their sons who live in the same house or in the same cluster of homes with them. In fact, the lack of challenges to mothers' definitions of time and tasks for their sons exhibits a cross-gender cohesion not found in other forms of gender relations.

## Conclusion

When one examines how Anamatians employ ideas of time and how they use discourse about time in social relationships, one sees that it is an important vehicle for the expression, performance, and maintenance of social differences. This chapter has discussed social difference with regard to the issues of class, ethnicity, age, and gender. In each case, temporal conceptions have been used to define difference and even to instigate open conflict. The public performance of ideas of time marks one's relationship to other groups and is most explicitly seen with regard to age conflicts and ethnic stereotypes. Time is not simply a dimension along which social activity is coordinated, but a means for differentiating groups in day-to-day practice.

# Conclusion

7

In his criticism of the contrast between "Western" linear ideas of time and "Oriental" cyclical ideas of time, Akhil Gupta suggests that "we need to ask why discursively available representations of time in the West remain oblivious, despite easily observable evidence to the contrary, to features of cyclicality, concreteness, rhythms, and yes, even rebirth" (1994:169). Gupta also points out that the dichotomy of the ideas of time found in industrial and non-industrial societies is equally difficult to sustain. In this contrast, industrial time is seen as "homogenous, empty, and regular" and non-industrial time is viewed as "rhythmic and irregular" (1994:172); yet, again, there is evidence of both times coexisting in many societies. While Gupta's intent is to challenge these notions of time as ideological constructions manifested in discourse about time, my intent has been to explore how this complicated coexistence of times affects the lives of people in Trinidad, a place that uneasily straddles the contrasts between East and West and between the industrial and the agricultural. This unease is the product of the intertwined histories of production and immigration in Trinidad.

The result of these historical and contemporary processes is that different ideas of time are distributed across age cohorts, institutions, and social roles. As shown in chapter 1, different generations have had different labor experiences. As a result, they have different views of the definitions of time used in modeling the relationship between employer and employee. In this way, experiences in the past affect relationships in the present. Sometimes these past experiences generate open contemporary conflict, as a result of differing conceptions of time. Chapter 2 discussed how different occupations involve different social rhythms. A consequent challenge for many workers is coordinating their work rhythms with those of others, such as the problems that emerge in the relationship between farmers and those who transport produce. Chapter 3 showed that ideas of time are distributed over social roles and that this, too, raises problems of temporal difference. Chapter 4 addressed institutionalized definitions of time and individual responses to these definitions. Ideas of time are distributed in a variety of ways, and this creates complications for the unfolding of social relationships.

Chapters 5 and 6 demonstrated that the multiplicity of times is a resource in claims of power, moral superiority, and social classifications. The idioms "jus' now," "any time is Trinidad time," and "long time" play an important role in managing temporal differences. They can be used to try to overcome such differences, or they can be used to exacerbate them. Importantly, Trinidadians use the manipulation of temporal differences to manipulate other social categories such as ethnicity, age, class, and gender. Ideas of time are also used in claims of power and challenges to such claims.

Consequently, in examining Trinidadian times, once one moves beyond the essentializing discourses that Gupta challenges, to look at the ways in which time is defined and the ways in which these temporal definitions are used, one realizes the culturally constructed and contested nature of time and the implications of these constructions and disputes for a wide range of social relationships.

While this study focuses on Trinidad, it implies the role played by cultural conceptions of time in all societies. Amid debates over relativism, the need for all societies to define time is a human universal, although the manner in which it is done may differ. Even differences are limited by natural factors. The recognition of day versus night is undoubtedly a human universal. Around this, the human body develops a rhythm of sleeping and wakefulness. Whatever this rhythm is, social life must be arranged around it. Other organisms have their own diurnal and seasonal cycles. Many societies pay attention to solar and lunar cycles (Hubert 1909; Isambert 1979; Nilsson 1920). Some societies also take into account the rhythms of the organisms they use and rely on when developing social rhythms as demonstrated by Evans-Pritchard (1940), Mauss (1979), and Hallowell (1955). Still other societies elaborate on these cycles with ritual and commercial cycles (Geertz 1973; Hoskins 1993; Howe 1981). The emphases that are placed on these events differ, however. Industrial discourses attempt to ignore natural times, such as the phases of the moon, and clocks are used to calibrate the sun, in the form of sunrise and sunset times, rather than the other way around.

Within these constraints, there is a great deal of variation. "Natural" time, the time of organisms with which people interact, is not universally consistent, nor is it ignored in so-called industrial societies. Cultural conceptions of time arising from Nuer pastoralism are undoubtedly different from rhythms arising from hunting and gathering, horticulture, or the cultivation of cocoa. In industrialized societies, time is often defined by human activities disengaged from natural phenomena, and it is occasionally defined by human bodily requirements of rest. Since the invention of the electric light, work is not limited to daytime hours. The demands of increasing production have generated night shifts. In these cases, work is determined by industrial rhythms expressed in the form of clock time. In popular imagery, ideologies of vocationally specific time have become a major dimension of defining certain professions. For example, we speak of bankers' hours and describe doctors as golfing on Wednesdays.

In effect, cultural conceptions of time arise from demands of production or from natural rhythms, and, in practice, they are influenced by the memories of the individuals involved. As societies change and grow more complex and specialization increases, the variety of times also increases across specialized occupations and across generations with different experiences. This creates strain on social organization. The examples from Anamat show that compromise and conflict are as important as power. Indeed, the strain on social coordination posed by multiple times is met through a variety of strategies for defining time and for overlooking temporal differences. These strategies provide resources for downplaying or marking differences between people. This is exemplified by the Trinidadian case.

What makes Trinidad interesting is the convergence of various forms of social differentiation—ethnicity, class, age, gender, and production. Since Anamat has participated in a variety of economic formations, each

defining time differently according to its demands for production, workers in Trinidad have contended with a variety of social definitions of time.
Thus, social life in Trinidad involves a nexus of various kinds of diversity:
ethnic diversity, occupational and vocational diversity, and, embedded in
age categories, diversity of historical experience. These factors interact. In
Anamat, there is a constant interplay between the image of the village as
being one people and the image of the village as consisting of many kinds
of people, of imposing order and generating disorder. These conflicting
ideologies are played out in daily life. In this context, the use of definitions
of time becomes an important medium for expressing cohesion, group
membership, difference, and antagonism. The nature of complexity in the
Anamatian situation highlights these issues—issues that arise in most, if
not all, complex societies.

In all societies, time is a medium for expressing social differences and
similarities both subtle and obvious. The more complex the social organization, the more important time as a medium of such expressions becomes. Time becomes more than simply a dimension in which social
activity is coordinated; it becomes the way in which people reflect upon
and represent temporal coordination. For instance, to be "fashionably
late" means, in one sense, to be late to a social engagement, but in another
sense, it highlights the fact that being late is fashionable and separates one
from others who are less fashionable by being on time. In some organizations, meetings start punctually, and those who arrive late are displayed as
marginal. In other organizations, the opposite is true—those who arrive
on time are shown to be marginal members, whereas the core members
always arrive late. Punctuality becomes locally defined, and interpretations having to do with belonging are linked to being punctual.

Multiple times and multiple social rhythms also encourage conflict and
tensions. The clock has become a means of mediating between people
who follow different temporal rhythms. But the clock is also used to define
and generate conflict. By means of the clock, appointments and schedules
are made, but because of the clock, it is obvious when appointments and
schedules are broken.

In the social sciences, conceptions of time have been viewed as necessary for the coordination of social activities. Indeed, it was with this assumption that I began. What has emerged from my ethnographic discussion of time in Trinidad is that while, in theory, such is the case, cultural

conceptions of time are also resources for the expression of many key facets of social organization: social differences such as ethnicity and gender, the exercise of power, and the expression of resistance. Time is not merely the means of coordinating social activities, although an ideology that it does so serves people well. It is also a means of keeping certain sets of relationships perpetually uncoordinated, as well as a means of controlling others. Social structure in complex societies emerges, in part, from the use of cultural conceptions of time. Since time is a resource to which, theoretically, everyone has access, it is a resource over which there is frequent interpretation, negotiation, and confrontation.

The problems Trinidadian society poses to social anthropological theory are those of any complex society. Factors such as class, occupational diversity, and ethnicity, as well as kinship, age, and gender, play significant roles in social organization. The use of conceptions of time and the use of strategies to highlight or obscure temporal differences help to define the social structural relationships between Trinidad's social categories. In part, this is because temporal conflicts diffuse through social networks, just as economic disputes diffuse through financial networks. If two individuals follow different models of time, and their relationships are not synchronized with another because of this, then this lack of synchronization affects other relationships. For instance, when doctors get behind in their schedules early in a day, they are delayed in fulfilling the remainder of their appointments. Those holding such appointments may, themselves, then be delayed in fulfilling their subsequent social engagements. Consequently, because time is a fundamental component of social organization, the interaction of two individuals can influence large social networks. Temporal chaos is contagious.

If the temporal difference is met with some means of obscuring the difference, such as tokens, glosses, or changing temporal models (see chapter 5), then the cohesion of the social structural links in the relationship are emphasized. If the difference is highlighted as a source of conflict, or is even generated to demonstrate difference, then the antagonisms in the social structural links are emphasized (see chapter 6). Whatever the result of the interaction with regard to negotiating temporal difference, the result is called into play in all interactions disrupted by the original temporal difference. For instance, temporal components of racial stereotypes may be expressed between an Indian and a Creole. These two individuals then might be delayed in meeting another obligation, at which time they account for their delay by presenting the stereotype.

Since social networks overlap and intersect, and since the effects of individual uses of temporal conceptions travel quickly through these networks, the relationships between social structural categories begin to emerge from the diffusion of the effects of individual interactions. On this basis, social actors develop a general conception of the configuration of social structural relationships. Indeed, conscious of such arrangements, some actors may deliberately attempt to overcome and change the relationships, say, using a gloss when conflict would normally be instigated and reinforced. This, in turn, diffuses through the networks of social relationships, and, given enough support through similar actions, it might even change the relationships between social structural categories. So, just as time reflects social structure, its manipulation can bring about change.

# Notes

### Introduction

1. This is probably one reason for the popularity of "Minute Rice," which is marketed as cooking perfectly every time.
2. See Freilich 1960a, 1960b, 1961, 1968, 1980, 1991, and Freilich and Coser 1972.
3. I use "European" to refer to anyone of European descent, including North Americans and Trinidadians of such descent.

### 2. Producing Times

1. In order to get up early, gardeners use a variety of methods ranging from having friends wake them up, to waking up to the sounds of animals stirring (dogs and roosters), to waking up to the sounds of increased traffic on the road, to using alarm clocks. With regard to the last, in the homes in which I have stayed, the alarm is turned off, then people return to sleep, rather than use it to wake up. In effect, the means that people use to wake up are as varied as the times that they construct out of their daily routines.
2. This disease is thought to be caused by a fungus that thrives in areas of poor drainage.
3. The United National Congress government, elected in 1995, placed temporary work in the Unemployment Relief Program under the Ministry of Local Government.

### 3. Distributed Times

1. Those who receive pensions from the government and those who work for the government are able to obtain their checks in Viego Grande on the first day of the month.
2. Parts of this section are based on Birth 1996.
3. See Brody (1989); DeBouzy (1979); O'Malley (1990); E. P. Thompson (1967); and Thrift (1981) for accounts of these reactions. It should also be noted that related to these complaints were concerns about the association workers made between wage labor and de-skilling.

### 4. Institutional and Consensual Temporal Coordination

1. The Department of Public Works under the colonial administration has now become the Ministry of Works, Decentralization, and Infrastructure.
2. Soca is a form of calypso which is popular at fetes. Its etymology is the combination of the words "soul" and "calypso." Kaiso is typically associated with slower tempos than soca, and its emphasis is on lyrics and wordplay rather than on melody and suitability for dancing.

### 5. Cohesion in Chaos

1. In 1970, amidst Black Power unrest, the army mutinied but was thwarted in an attempt to overthrow the government due to swift action by the coast guard and the police. For more information on this event, see Pantin (1990), Owen Baptiste (1976), Ryan (1972), Ryan and Stewart (1995), and Oxaal (1982).
2. On July 27, 1990, there was an attempted overthrow of the government. This interview occurred approximately six months after the coup attempt.

### 6. Social Conflict and Time

1. There are no tailor shops in Anamat, but there are several Anamatian tailors who work outside Anamat. Their shops are important nodes in information networks connecting Anamat to other parts of Trinidad. Consequently, they are worthwhile places to lime for those who seek the latest news.
2. "Wife" is used here as a general term to include any co-residential conjugal union, including those unions not legally recognized or recognized by some religious organization.

# Bibliography

Abdool, Kamal

1979   P-H Transport. St. Augustine: University of the West Indies, Department of Sociology.

Abrahams, Roger

1983   The Man of Words in the West Indies. Baltimore: Johns Hopkins University Press.

Adam, Barbara

1995   Timewatch: The Social Analysis of Time. Cambridge, Mass.: Polity Press.

Agricultural Policy Committee of Trinidad and Tobago

1943   Report. Port of Spain: A. L. Rhodes, Government Printer.

Anglo-American Caribbean Commission

1943   The Caribbean Islands and the War: A Record of Progress in Facing Stern Realities. Washington, D.C.: U.S. Government Printing Office.

Angrosino, Michael V.

1974   Outside Is Death. Winston-Salem, N.C.: Overseas Research Center.

Anthony, Michael

1989   Parade of the Carnivals of Trinidad, 1839–1989. St. James, Port of Spain: Circle Press.

Appadurai, Arjun

1996   Modernity at Large: Cultural Dimensions of Globalization. Minneapolis: University of Minnesota Press.

Baksh, Ishmael

1979   Stereotypes of Negroes and East Indians in Trinidad: A Re-examination. Caribbean Quarterly 25:52–71.

Baptiste, Fitzroy

1988   War, Cooperation, and Conflict: The European Possessions in the Caribbean, 1939–1945. New York: Greenwood.

Baptiste, Owen, ed.

1976  *Crisis.* St. James, Trinidad: Inprint.

Berleant-Schiller, Riva, and Lydia M. Pulsipher

1986  Subsistence Cultivation in the Caribbean. *Nieuwe West-Indische Gids* 60:1–40.

Birth, Kevin

1990  Reading and the Righting of Writing Ethnographies. *American Ethnologist* 17:549–57.

1996  Trinidadian Times: Temporal Dependency and Temporal Flexibility on the Margins of Industrial Capitalism. *Anthropological Quarterly* 69:79–89.

Blanshard, Paul

1947  *Democracy and Empire in the Caribbean.* New York: Macmillan.

Bloch, Maurice

1977  The Past and the Present in the Present. *Man* 12:278–92.

1979  Knowing the World or Hiding It. *Man* 14:165–67.

Blouet, W. B.

1976  The Post-Emancipation Origins of the Relationships between the Estates and the Peasantry in Trinidad. In *Land and Labour in Latin America*, ed. K. Duncan and I. Rutledge, 435–51. Cambridge: Cambridge University Press.

Borst, Arno

1993  *The Ordering of Time: From the Ancient Computus to the Modern Computer*, trans. Andrew Winnard. Chicago: University of Chicago Press.

Bourdieu, Pierre

1963  The Attitude of the Algerian Peasant toward Time. In *Mediterranean Countrymen*, ed. Julian Pitt-Rivers, 55–72. The Hague: Mouton.

1979  The Disenchantment of the World. In *Algeria 1960*, trans. Richard Nice, 1–94. Cambridge: Cambridge University Press.

Bourdillon, M. F. C.

1978  Knowing the World or Hiding It: A Response to Maurice Bloch. *Man* 13:591–99.

Brana-Shute, Gary

1976  Drinking Shops and Social Structure: Some Ideas of Lower Class West Indian Male Behavior. *Urban Anthropology* 5:53–68.

1979  *On the Corner: Male Social Life in a Paramaribo Creole Neighborhood.* Assen, Netherlands: Van Gorcum.

Brereton, Bridget

1974  The Foundations of Prejudice: Indians and Africans in Nineteenth Century Trinidad. *Caribbean Issues* 1:15–28.

1979  *Race Relations in Colonial Trinidad, 1870–1900.* London: Cambridge University Press.

1981  *A History of Modern Trinidad, 1783–1962.* Kingston: Heinemann.

1985 (1974)The Experience of Indentureship: 1845–1917. In *Calcutta to Caroni*, ed. John La Guerre, 25–38. St. Augustine: University of the West Indies, Extra-Mural Studies Unit.

1992 Ex-Slaves, Villagers, Squatting, Agriculture, 1838–1900. In *The Book of Trinidad*, ed. Gerard Besson and Bridget Brereton, 173–74. Port of Spain: Paria.

1993 Social Organisation and Class, Racial and Cultural Conflict in Nineteenth Century Trinidad. In *Trinidad Ethnicity*, ed. Kevin Yelvington, 33–55. Knoxville: University of Tennessee Press.

Brody, D.

1989 Time and Work during Early American Industrialism. *Labor History* 30:5–46.

Burnley, W. H.

1842 *Observations on the Present Condition of the Island of Trinidad*. London: Longman, Brown, Green, and Longmans.

Campbell, Carl

1992 *Colony and Nation: A Short History of Education in Trinidad and Tobago*. Kingston, Jamaica: Ian Randle.

Carmichael, Gertrude

1961 *The History of the West Indian Islands of Trinidad and Tobago*. London: Alvin Redman.

Carrington, Edwin

1971 The Post War Political Economy of Trinidad and Tobago. In *Readings in the Political Economy of the Caribbean*, ed. Norman Girvan and Owen Jefferson, 121–42. Kingston, Jamaica: New World Group.

Chapman, J. K.

1964 *The Career of Arthur Hamilton Gordon, First Lord Stanmore, 1829–1912*. Toronto: University of Toronto Press.

Clifford, James

1986 Introduction: Partial Truths. In *Writing Culture*. ed. James Clifford and George E. Marcus, 1–26. Berkeley: University of California Press.

Comitas, Lambros

1973 Occupational Multiplicity in Rural Jamaica. In *Work and Family Life: West Indian Perspectives*, ed. Lambros Comitas and David Lowenthal, 156–73. Garden City, N.Y.: Anchor.

Craig, Susan

1985 Political Patronage and Community Resistance: Village Councils in Trinidad and Tobago. In *Rural Development in the Caribbean*, ed. P. I. Gomes, 173–93. Kingston, Jamaica: Heinemann Educational Books (Caribbean).

Curtin, Philip

1954 The British Sugar Duties and West Indian Prosperity. *Journal of Economic History* 14:157–64.

de Boissiere, Jean

1992 The French in Trinidad. In *The Book of Trinidad*, ed. Gerard Besson and Bridget Brereton, 140–49. Port of Spain: Paria.

DeBouzy, Marianne

1979 Aspects du temps industriel aux États-Unis au debut du XIXe siècle. *Cahiers Internationaux de Sociologie* 57:197–220.

de Verteuil, Anthony

1984 *The Years of Revolt: Trinidad, 1881–1888*. Port of Spain: Paria.

1994 *The Germans of Trinidad*. Port of Spain: Litho.

Dirks, Robert

1987 *The Black Saturnalia: Conflict and Its Ritual Expression on British West Indian Slave Plantations*. Gainesville: University of Florida Press.

Dohrn-van Rossum, Gerhard

1996 *History of the Hour: Clocks and Modern Temporal Orders*, trans. Thomas Dunlap. Chicago: University of Chicago Press.

Durkheim, Emile

1965 (1915) *The Elementary Forms of the Religious Life*. New York: Free Press.

Eriksen, Thomas Hylland

1990 Liming in Trinidad: The Art of Doing Nothing. *Folk* 32:23–43.

Evans-Pritchard, E. E.

1940 *The Nuer: A Description of the Modes of Livelihood and Political Institutions of a Nilotic People*. Oxford: Oxford University Press.

Firth, Raymond

1964 *Essays on Social Organization and Values*. London: Athlone.

Foucault, Michel

1977 *Discipline and Punish: The Birth of the Prison*. London: Allen Lane.

Frankson, Geoffrey

1981 *D. R. Coldson Medical Opinion: Selected Essays, 1978–1980*. Port of Spain: Inprint.

Freilich, Morris

1960a Cultural Diversity among Trinidadian Peasants. Ph.D. dissertation, Columbia University.

1960b Cultural Models and Land Holdings. *Anthropological Quarterly* 33:188–97.

1961 Serial Polygyny, Negro Peasants, and Modal Analysis. *American Anthropologist* 63:955–75.

1968 Sex, Secrets, and Systems. In *The Family in the Caribbean*, ed. Stanford Gerber, 47–62. Rio Piedras: Institution of Caribbean Studies.

1980 Smart-Sex and Proper-Sex. *Central Issues in Anthropology* 2:37–51.

1991 Smart Rules and Proper Rules: A Journey through Deviance. In *Deviance: Anthropological Perspectives*, ed. Morris Freilich, Douglas Raybeck, and Joel Savishinsky, 27–50. New York: Bergin and Garvey.

Freilich, Morris, and Lewis A. Coser

1972 Structured Imbalances of Gratification: The Case of the Caribbean Mating System. *British Journal of Sociology* 23:1−19.

Geertz, Clifford

1973 Person, Time, and Conduct in Bali. In *The Interpretation of Cultures,* 360−411. New York: Basic Books.

Gell, Alfred

1992 *The Anthropology of Time: Cultural Constructions of Temporal Maps and Images.* Oxford: Berg.

Genovese, Eugene

1976 *Roll Jordan Roll: The World the Slaves Made.* New York: Random House.

Gilbert, S. M.

1931 *The Cocoa Industry of Trinidad.* Council Paper no. 4 of 1931. Port of Spain: Government Printing Office.

Glasser, Richard

1972 *Time in French Life and Thought,* trans. C. G. Pearson. Manchester: Manchester University Press.

Glassie, Henry H.

1982 *Passing the Time in Ballymenone.* Philadelphia: University of Pennsylvania Press.

Goffman, Erving

1959 *The Presentation of Self in Everyday Life.* Garden City, N.Y.: Anchor Books.

1967 On Face Work. In *Interaction Ritual: Essays on Face-to-Face Behavior,* 5−45. New York: Pantheon Books.

Goodenough, Ward H.

1981 (1973) *Culture, Language, and Society.* Menlo Park, Calif.: Benjamin/Cummings.

Greenhouse, Carol

1996 *A Moment's Notice: Time Politics across Cultures.* Ithaca, N.Y.: Cornell University Press.

Gupta, Akhil

1994 The Reincarnation of Souls and the Rebirth of Commodities: Representations of Time in "East" and "West." In *Remapping Memory: The Politics of TimeSpace,* ed. Jonathan Boyarin, 161−83. Minneapolis: University of Minnesota Press.

Gutman, Herbert

1976 *Work, Culture, and Society in Industrializing America: Essays in American Working-Class and Social History.* New York: Alfred A. Knopf.

Hall, Edward T.

1983 *The Dance of Life.* Garden City, N.Y.: Anchor.

Hallowell, A. Irving

1937 (1955) Temporal Orientation in Western Civilization and in a Preliterate Society. In *Culture and Experience,* 216−35. Philadelphia: University of Pennsylvania Press.

Hannerz, Ulf

1992   *Cultural Complexity: Studies in the Social Organization of Meaning.* New York: Columbia University Press.

Hareven, Tamara K.

1982   *Family Time and Industrial Time: The Relationship between the Family and Work in a New England Industrial Community.* Cambridge: Cambridge University Press.

Harry, Indra S.

1993   Women in Agriculture in Trinidad: An Overview. In *Women and Change in the Caribbean,* ed. Janet H. Momsen, 194–204. Indiana: Indiana University Press.

Henry, Frances, and Pamela Wilson

1975   The Status of Women in Caribbean Societies: An Overview of Their Social, Economic, and Sexual Roles. *Social and Economic Studies* 24: 165–98.

Henry, Ralph

1988   The State and Income Distribution in an Independent Trinidad and Tobago. In *Trinidad and Tobago: The Independence Experience, 1962–1987,* ed. Selwyn Ryan, 471–93. St. Augustine, Trinidad: Institute of Social and Economic Research.

Higman, B. W.

1984   *Slave Populations of the British Caribbean 1807–1834.* Baltimore: Johns Hopkins University Press.

Hill, Polly

1986   *Development Economics on Trial.* Cambridge: Cambridge University Press.

Hills, Theo L.

1988   The Caribbean Peasant Food Forest, Ecological Artistry or Random Chaos. In *Small Farming and Peasant Resources in the Caribbean,* ed. John S. Brierley and Hymie Rubenstein, 1–28. Winnipeg, Canada: University of Manitoba, Department of Geography.

Hintzen, Percy C.

1989   *The Costs of Regime Survival: Racial Mobilization, Elite Domination, and Control of the State in Guyana and Trinidad.* Cambridge: Cambridge University Press.

Ho, Christine

1991   *Salt-Water Trinnies: Afro-Trinidadian Immigrant Networks and Non-Assimilation in Los Angeles.* New York: AMS.

Hoskins, Janet

1993   *The Play of Time: Kodi Perspectives on Calendars, History, and Exchange.* Berkeley: University of California Press.

Howe, Leopold E. A.

1981   The Social Determination of Knowledge: Maurice Bloch and Balinese Time. *Man* 16: 220–34.

Hubert, Henri

1909   Étude sommaire de la représentation du temps dans la religion et la magie.

In *Mélanges d'histoire des religions*, H. Hubert and M. Mauss, 189–229. Paris: Félix Alcan.

Isambert, François A.

1979 Henri Hubert et la Sociologie du Temps. *Revue Française de Sociologie* 20: 183–201.

James, C. L. R.

1993 *Beyond a Boundary*. Durham, N.C.: Duke University Press.

Johnson, Howard

1971 Immigration and the Sugar Industry in Trinidad during the Last Quarter of the 19th Century. *Journal of Caribbean History* 3: 28–72.

Johnston, Robert A., and James C. Shoultz

1945–47 *History of the Trinidad Sector and Base Command*. Washington, D.C.: U.S. Army Caribbean Defense Command.

Jolly, A. L.

1954a *Peasant Farming: Report on Peasant Experimental Farms at the Imperial College of Tropical Agriculture, Trinidad, B.W.I.* Port of Spain: Caribbean Commission.

1954b Small Scale Farm Management Problems. In *Small Scale Farming in the Caribbean*, Conference on Education and Small Scale Farming, 15–24. Port of Spain: Caribbean Commission.

Kant, Immanuel

1929 *Critique of Pure Reason*, trans. Norman Kemp Smith. New York: St. Martin's Press.

Khan, Aisha

1994 Juthaa in Trinidad: Food, Pollution, and Hierarchy in a Caribbean Diaspora Community. *American Ethnologist* 21: 245–69.

Klass, Morton

1961 *East Indians in Trinidad*. New York: Columbia University Press.

1995 (1991) *Singing with Sai Baba: The Politics of Revitalization in Trinidad*. Prospect Heights, Ill.: Waveland.

La Guerre, John Gaffar

1972 The General Elections of 1946 in Trinidad and Tobago. *Social and Economic Studies* 21: 184–204.

Lakoff, George, and Mark Johnson

1980 *Metaphors We Live By*. Chicago: University of Chicago Press.

Landes, David

1983 *Revolution in Time: Clocks and the Making of the Modern World*. Cambridge, Mass.: Belknap.

Laurence, K. O.

1971 *Immigration into the West Indies in the 19th Century*. St. Lawrence, Barbados: Caribbean Universities Press.

1994 *A Question of Labour: Indentured Immigration into Trinidad and British Guiana, 1875–1917*. New York: St. Martin's Press.

Lazarus-Black, Mindie

1994   *Legitimate Acts and Illegal Encounters: Law and Society in Antigua and Barbuda.* Washington, D.C.: Smithsonian Institution Press.

Leclerq, Jean

1975   Experience and the Interpretation of Time in the Early Middle Ages. *Studies in Medieval Culture* 5:9–19.

Le Goff, Jacques

1980   *Time, Work, and Culture in the Middle Ages,* trans. Arthur Goldhammer. Chicago: University of Chicago Press.

Lewis, Gordon K.

1968   *The Growth of the Modern West Indies.* New York: Monthly Review Press.

Lieber, Michael

1976   "Liming" and Other Concerns: The Style of Street Embedments in Port of Spain, Trinidad. *Urban Anthropology* 5:319–34.

Lobdell, Richard

1972   Patterns of Investment and Credit in the British West Indian Sugar Industry, 1838–1897. *Journal of Caribbean History* 4:31–53.

Look Lai, Walton

1993   *Indentured Labor, Caribbean Sugar: Chinese and Indian Migrants to the British West Indies, 1838–1918.* Baltimore: Johns Hopkins University Press.

Marcus, George, and Michael M. J. Fischer

1986   *Anthropology as Cultural Critique.* Chicago: University of Chicago Press.

Marsden, E. J.

1945   Trinidad's War Effort. *Canada-West Indies Magazine* 35:29, 31, 33.

Marx, Karl

1977   *Capital,* trans. Ben Fowkes. Vol. 1. New York: Vintage.

Mauss, Marcel

1979   *Seasonal Variation among the Eskimo: A Study in Social Morphology,* trans. James Fox. London: Routledge and Kegan Paul.

Miller, Daniel

1991   Absolute Freedom in Trinidad. *Man* 27:323–41.

1994   *Modernity—An Ethnographic Approach: Dualism and Mass Consumption in Trinidad.* Oxford: Berg.

1995   Introduction: Anthropology, Modernity and Consumption. In *Worlds Apart: Modernity through the Prism of the Local,* ed. Daniel Miller, 1–22. New York: Routledge.

1997   *Capitalism: An Ethnographic Approach.* Oxford: Berg.

Mintz, Sidney W.

1961   The Question of Caribbean Peasantries: A Comment. *Caribbean Studies* 1:31–34.

1973  A Note on the Definition of Peasantry. *Journal of Peasant Studies* 1:91–106.

1974  *Caribbean Transformations*. Chicago: Aldine.

1979  Slavery and the Rise of the Peasantry. *Historical Reflections* 6:215–42.

1983  Reflections on Caribbean Peasantries. *Nieuwe West-Indische Gids* 57:1–18.

1985a From Plantations to Peasantries in the Caribbean. In *Caribbean Contours*, ed. Sidney Mintz and Sally Price, 127–54. Baltimore: Johns Hopkins University Press.

1985b *Sweetness and Power: The Place of Sugar in Modern History*. New York: Penguin.

Mischel, Walter

1958  Preference for Delayed Reinforcement: An Experimental Study of a Cultural Observation. *Journal of Abnormal and Social Psychology* 56:57–61.

1961a Preference for Delayed Reinforcement and Social Responsibility. *Journal of Abnormal and Social Psychology* 62:1–7.

1961b Delay of Gratification, Need for Achievement, and Acquiescence in Another Culture. *Journal of Abnormal and Social Psychology* 62:543–52.

1961c Father-Absence and Delay of Gratification: Cross-Cultural Comparisons. *Journal of Abnormal and Social Psychology* 63:116–24.

Moodie-Kublalsingh, Sylvia

1994  *The Cocoa Panyols of Trinidad: An Oral Record*. London: British Academic Press.

Moore, W. E.

1963a The Temporal Structure of Organizations. In *Sociological Theory, Values and Sociocultural Change*, ed. E. A. Tiryakian, 161–69. London: Free Press.

1963b *Man, Time and Society*. New York: Wiley.

Mumford, Lewis

1963 (1934)  *Technics and Civilization*. New York: Harcourt, Brace, and World.

Munn, Nancy

1983  Gawan Jula: Spatiotemporal Control and the Symbolism of Influence. In *The Kula: New Perspectives on Massim Exchange*, ed. Jerry W. Leach and Edmund Leach, 277–308. Cambridge: Cambridge University Press.

1992  The Cultural Anthropology of Time: A Critical Essay. *Annual Review of Anthropology* 21:93–123.

Naipaul, V. S.

1981 (1962)  *The Middle Passage*. New York: Vintage.

Newson, L. A.

1976  *Aboriginal and Spanish Colonial Trinidad*. London: Academic Press.

Niehoff, Arthur, and Janita Neihoff

1960  *East Indians in the West Indies*. Milwaukee: Milwaukee Public Museum Publications in Anthropology, no. 6.

Nilsson, Martin P.

1920    *Primitive Time-Reckoning: A Study in the Origins and First Development of the Art of Counting Time among the Primitive Early Culture Peoples.* Lund: C. W. K. Gleerup.

Olwig, Karen Fog

1993    *Global Culture, Island Identity: Continuity and Change in the Afro-Caribbean Community of Nevis.* Chur, Switzerland: Harewood Academic Publishers.

O'Malley, Michael

1990    *Keeping Watch: A History of American Time.* New York: Viking.

1992    Time, Work and Task Orientation. *Time and Society* 1:341–58.

Östör, Akos

1993    *Vessels of Time.* New York: Oxford University Press.

Oxaal, Ivar

1982    *Black Intellectuals and the Dilemmas of Race and Class in Trinidad.* Rochester, Vt.: Schenkman.

Pantin, Raoul

1990    *Black Power Day: The 1970 February Revolution.* Santa Cruz, Trinidad: Hatuey Productions.

Pares, Richard

1956    *Yankees and Creoles.* Cambridge: Harvard University Press.

Phillips-Lewis, Kathleen E.

1988    Peasant Cocoa Cultivators in Trinidad, 1870–1920: Some Considerations. In *Small Farming and Peasant Resources in the Caribbean,* ed. John S. Brierley and Hymie Rubenstein, 29–38. Winnipeg, Canada: University of Manitoba, Department of Geography.

Premdas, Ralph

1993    Race, Politics, and Succession in Trinidad and Guyana. In *Modern Caribbean Politics,* ed. Anthony Payne and Paul Sutton, 98–124. Baltimore: Johns Hopkins University Press.

Ramsaran, Ramesh F.

1994    The Theory and Practice of Structural Adjustment with Specific Reference to the Commonwealth Caribbean. In *Structural Adjustment, Public Policy and Administration in the Caribbean,* ed. John La Guerre, 9–37. St. Augustine, Trinidad: University of the West Indies, School of Continuing Studies.

Reddock, Rhoda

1993    Transformation in the Needle Trades: Women in Garment and Textile Production in Early Twentieth-Century Trinidad. In *Women and Change in the Caribbean,* ed. Janet H. Momsen, 249–62. Bloomington: Indiana University Press.

1994    *Women, Labour and Politics in Trinidad and Tobago: A History.* London: Zed.

Richardson, Bonham

1992   *The Caribbean in the Wider World, 1492–1992.* Cambridge: Cambridge University Press.

Riviere, W. E.

1972   Labour Shortage in the British West Indies after Emancipation. *Journal of Caribbean History* 4: 1–30.

Rutz, Henry J.

1992   Introduction: The Idea of a Politics of Time. In *The Politics of Time,* ed. Henry J. Rutz, 1–17. Washington, D.C.: American Ethnological Monograph Series, no. 4.

Rutz, Henry, and Erol M. Balkan

1992   Never on Sunday: Time-Discipline and Fijian Nationalism. In *The Politics of Time,* ed. Henry J. Rutz, 62–85. Washington, D.C.: American Ethnological Monograph Series, no. 4.

Ryan, Selwyn

1972   *Race and Nationalism in Trinidad and Tobago.* Toronto: University of Toronto Press.

1989a  *Revolution and Reaction: Parties and Politics in Trinidad and Tobago, 1970–1981.* St. Augustine, Trinidad: Institute of Social and Economic Research.

1989b  *The Disillusioned Electorate: The Politics of Succession in Trinidad and Tobago.* Port of Spain: Inprint.

1991a  *The Muslimeen Grab for Power: Race, Religion and Revolution in Trinidad and Tobago.* Port of Spain: Inprint.

1991b  Race and Occupational Stratification in Trinidad and Tobago. In *Social and Occupational Stratification in Contemporary Trinidad and Tobago,* ed. Selwyn Ryan, 166–90. St. Augustine, Trinidad: Institute of Social and Economic Research.

Ryan, Selwyn, Eddie Greene, and Jack Harewood

1979   *The Confused Electorate: A Study of Political Attitudes and Opinions in Trinidad and Tobago.* St. Augustine, Trinidad: Institute of Social and Economic Research.

Ryan, Selwyn, and Taimoon Stewart, eds.

1995   *Power: The Black Power Revolution, 1970.* St. Augustine, Trinidad: Institute of Social and Economic Research.

Safa, Helen

1995   *The Myth of the Male Breadwinner: Women and Industrialization in the Caribbean.* Boulder, Col.: Westview Press.

Samaroo, Brinsley

1985 (1974) Politics and Afro-Indian Relations in Trinidad. In *Calcutta to Caroni,* ed. John La Guerre, 77–92. St. Augustine: University of the West Indies, Extra-Mural Unit.

Sauer, Carl

1966   *The Early Spanish Main.* Berkeley: University of California Press.

Schwartz, Barton

1964a Caste and Endogamy in Trinidad. *Southwestern Journal of Anthropology* 20:58–66.

1964b Ritual Aspects of Caste in Trinidad. *Anthropological Quarterly* 37:1–15.

1967 The Failure of Caste in Trinidad. In *Caste in Overseas Indian Communities*, ed. Barton Schwartz, 213–55. San Francisco: Chandler.

Schwartz, Theodore

1978 Where Is the Culture? Personality as the Distributive Locus of Culture. In *The Making of Psychological Anthropology*, ed. George D. Spindler, 419–41. Berkeley: University of California Press.

Searle, Chris

1991 The Muslimeen Insurrection in Trinidad. *Race and Class* 33:29–43.

Segal, Daniel A.

1993 "Race" and "Colour" in Pre-Independence Trinidad and Tobago. In *Trinidad Ethnicity*, ed. Kevin Yelvington, 81–115. Knoxville: University of Tennessee Press.

Senior, Olive

1991 *Working Miracles: Women's Lives in the English-Speaking Caribbean.* Bloomington: Indiana University Press.

Sewell, William

1968 (1861) *The Ordeal of Free Labour in the British West Indies.* New York: Reprints of Economic Classics.

Shephard, C. Y.

1932 The Cacao Industry of Trinidad: Some Economic Aspects. *Tropical Agriculture* 9:95–100, 145–52, 185–95, 200–205, 236–43, 307–17, 334–45.

1935 Agricultural Labour in Trinidad. *Tropical Agriculture* 12:3–9, 43–47, 56–64, 84–88, 126–31, 153–57, 187–92.

1936 *The Cocoa Industry of Trinidad, Series II: A Financial Survey of Estates during the Seven Years* 1923–24 to 1929–30. Port of Spain: Government Printing Office.

1954 Organisation for the Processing and Marketing of the Products of Small Scale Farming. In *Small Scale Farming in the Caribbean*, Conference on Education and Small Scale Farming, 40–52. Port of Spain: Caribbean Commission.

Shore, Bradd

1996 *Culture in Mind: Cognition, Culture, and the Problem of Meaning.* New York: Oxford University Press.

Singh, Kelvin

1985 (1974) Indians and the Larger Society. In *Calcutta to Caroni: The East Indians of Trinidad*, ed. John La Guerre, 33–60. St. Augustine: University of the West Indies, Extra-Mural Studies Unit.

1994 *Race and Class Struggles in a Colonial State: Trinidad, 1917–1945.* Mona, Jamaica: The Press—University of the West Indies.

Smith, Mark M.

1996    Old South Time in Comparative Perspective. *The American Historical Review* 101:1432–69.

1997    *Mastered by the Clock*. Chapel Hill: University of North Carolina Press.

Smith, Michael French

1982    Bloody Time and Bloody Scarcity. *American Ethnologist* 9:503–18.

Smith, R. T.

1959    Some Social Characteristics of Indian Immigrants to British Guiana. *Population Studies* 13:34–39.

1996    *The Matrifocal Family*. New York: Routledge.

Sorokin, Pitirim

1964    *Sociocultural Causality, Space, Time: A Study of Referential Principles of Sociology and Social Science*. New York: Russell and Russell.

Sorokin, Pitirim A., and Robert K. Merton

1937    Social Time: A Methodological and Functional Analysis. *American Journal of Sociology* 42:615–29.

Stewart, John O.

1986    Patronage and Control in the Trinidadian Carnival. In *The Anthropology of Experience*, ed. Victor Turner and Edward Bruner, 289–315. Urbana: University of Illinois Press.

Stuempfle, Stephen

1995    *The Steelband Movement: The Forging of a National Art in Trinidad and Tobago*. Philadelphia: University of Pennsylvania Press.

Sutton, Paul

1984    Trinidad and Tobago: Oil Capitalism and the "Presidential Power" of Eric Williams. In *Dependency under Challenge: The Political Economy of the Commonwealth Caribbean*, ed. Anthony Payne and Paul Sutton, 43–76. Manchester: Manchester University Press.

Swartz, Marc J.

1984    Culture as Token and as Guides: Swahili Views and Behavior concerning Generational Differences. *Journal of Anthropological Research* 40:78–89.

1990    Aggressive Speech, Status, and Cultural Distribution among the Swahili of Mombasa. In *Personality and the Cultural Construction of Society: Essays in Honor of M. E. Spiro*, ed. D. K. Jordan and M. J. Swartz, 116–42. Tuscaloosa: University of Alabama Press.

1991    *The Way the World Is*. Berkeley: University of California Press.

Taussig, Michael T.

1980    *The Devil and Commodity Fetishism in South America*. Chapel Hill: University of North Carolina Press.

Thomas, Clive

1988   *The Poor and the Powerless: Economic Policy and Change in the Caribbean.* New York: Monthly Review Press.

Thompson, E. P.

1967   Time, Work-Discipline, and Industrial Capitalism. *Past and Present* 38:56–97.

Thrift, Nigel

1981   Owners' Time and Own Time: The Making of a Capitalist Time Consciousness, 1300–1880. In *Space and Time in Geography: Essays Dedicated to Tarsten Hägerstrand,* ed. Allan Pred, 56–84. Lund: Lund Studies in Geography, Series B, Human Geography no. 48.

1988   Vivos Voco: Ringing the Changes in the Historical Geography of Time Consciousness. In *The Rhythms of Society,* ed. Michael Young and Tom Schuller, 53–94. London: Routledge.

Trinidad Express Newspapers

1990   *Trinidad under Siege, the Muslimeen Uprising: Six Days of Terror.* Port of Spain: Trinidad Express Newspapers.

Trotman, David V.

1986   *Crime in Trinidad.* Knoxville: University of Tennessee Press.

Trouillot, Michel-Rolph

1988   *Peasants and Capital: Dominica in the World Economy.* Baltimore: Johns Hopkins University Press.

Verdery, Katherine

1992   The "Etatization" of Time in Ceausescu's Romania. In *The Politics of Time,* ed. Henry J. Rutz, 37–61. Washington, D.C.: American Ethnological Society Monograph Series, no. 4.

Waterman, James A.

1967   Malaria—and Its Eradication in Trinidad and Tobago. *Caribbean Medical Journal* 29:19–35.

Watts, David

1987   *The West Indies: Patterns of Development, Culture and Environmental Change since 1492.* Cambridge: Cambridge University Press.

Weber, Max

1976 (1958) *The Protestant Ethic and the Spirit of Capitalism.* New York: Charles Scribner's Sons.

West India Royal Commission

1945   *Report.* London: His Majesty's Stationery Office.

Whipp, Richard

1987   "A Time to Every Purpose": An Essay on Time and Work. In *The Historical Meanings of Work,* ed. Patrick Joyce, 216–36. Cambridge: Cambridge University Press.

Whitrow, G. J.

1988  *Time in History.* Oxford: Oxford University Press.

Williams, Brackette

1987  Humor, Linguistic Ambiguity and Disputing in a Guyanese Community. *International Journal of the Sociology of Language* 65:79–94.

1991  *Stains on My Name, War in My Veins: Guyana and the Politics of Cultural Struggle.* Durham, N.C.: Duke University Press.

Williams, Eric

1964 (1944) *Capitalism and Slavery.* London: Andre Deutsch.

Willis, Paul

1977  *Learning to Labor: How Working Class Kids Get Working Class Jobs.* New York: Columbia University Press.

Wilson, Peter J.

1969  Reputation and Respectability: A Suggestion for Caribbean Ethnology. *Man* 4:70–84.

1971  Caribbean Crews: Peer Groups and Male Society. *Caribbean Studies* 10:18–34.

1973  *Crab Antics.* New Haven: Yale University Press.

Wolf, Eric

1966  *Peasants.* Englewood Cliffs, N.J.: Prentice Hall.

1982  *Europe and the People without History.* Berkeley: University of California Press.

Wood, Donald

1968  *Trinidad in Transition: The Years after Slavery.* Oxford: Oxford University Press.

Yelvington, Kevin

1987  Vote Dem Out: The Demise of the PNM in Trinidad and Tobago. *Caribbean Review* 14:8–33.

1993a Introduction. In *Trinidad Ethnicity,* ed. Kevin Yelvington, 1–32. Knoxville: University of Tennessee Press.

1993b Gender and Ethnicity at Work in a Trinidadian Factory. In *Women and Change in the Caribbean,* ed. Janet H. Momsen, 263–77. Bloomington: Indiana University Press.

1995  *Producing Power: Ethnicity, Gender, and Class in a Caribbean Workplace.* Philadelphia: Temple University Press.

Zerubavel, Eviatar

1979  *Patterns of Time in Hospital Life.* Chicago: University of Chicago Press.

1981  *Hidden Rhythms: Schedules and Calendars in Social Life.* Chicago: University of Chicago Press.

# Index

Kevin K. Birth is associate profes-
sor of anthropology at Queens
College of the City University of
New York. He specializes in psychological anthropology, social
anthropology, and the Caribbean.
His articles have appeared in
*American Ethnologist, Ethnology,* and *Anthropological Quarterly.*